ANIMAL AND PLANT
Anatomy

VOLUME CONSULTANTS

- Barbara Abraham, *Hampton University, VA* • Amy-Jane Beer, *Natural history writer and consultant*
- Erica Bower, *Botanical writer and researcher* • Sally-Anne Mahoney, *Bristol University, England*
- Chris Mattison, *Natural history writer and researcher* • Kieran Pitts, *Bristol University, England*
- John Stewart, *Natural History Museum, London, England* • Phil Whitfield, *King's College, London*

2
Crocodile – Fern

 **Marshall Cavendish**
Reference
New York

CONTRIBUTORS

Roger Avery; Richard Beatty; Amy-Jane Beer; Erica Bower; Trevor Day; Erin Dolan; Bridget Giles; Natalie Goldstein; Tim Harris; Christer Hogstrand; Rob Houston; John Jackson; Tom Jackson; James Martin; Chris Mattison; Katie Parsons; Ray Perrins; Kieran Pitts; Adrian Seymour; Steven Swaby; John Woodward.

CONSULTANTS

Barbara Abraham, Hampton University, VA; Glen Alm, University of Guelph, Ontario, Canada; Roger Avery, Bristol University, England; Amy-Jane Beer, University of London, England; Deborah Bodolus, East Stroudsburg University, PA; Allan Bornstein, Southeast Missouri State University, MO; Erica Bower, University of London, England; John Cline, University of Guelph, Ontario, Canada; Trevor Day, University of Bath, England; John Friel, Cornell University, NY; Valerius Geist, University of Calgary, Alberta, Canada; John Gittleman, University of Virginia, VA; Tom Jenner, University of Cardiff, Wales; Bill Kleindl, University of Washington, Seattle, WA; Thomas Kunz, Boston University, MA; Alan Leonard, Florida Institute of Technology, FL; Sally-Anne Mahoney, Bristol University, England; Chris Mattison; Andrew Methven, Eastern Illinois University, IL; Graham Mitchell, King's College, London, England; Richard Mooi, California Academy of Sciences, San Francisco, CA; Ray Perrins, Bristol University, England; Kieran Pitts, Bristol University, England; Adrian Seymour, Bristol University, England; David Spooner, University of Wisconsin, WI; John Stewart, Natural History Museum, London, England; Erik Terdal, Northeastern State University, Broken Arrow, OK; Phil Whitfield, King's College, University of London, England.

Marshall Cavendish
99 White Plains Road
Tarrytown, NY 10591–9001

www.marshallcavendish.us

Library of Congress Cataloging-in-Publication Data
Animal and plant anatomy.
    p. cm.
  ISBN-13: 978-0-7614-7662-7 (set: alk. paper)
  ISBN-10: 0-7614-7662-8 (set: alk. paper)
  ISBN-13: 978-0-7614-7664-1 (vol. 2)
  ISBN-10: 0-7614-7664-4 (vol. 2)
  1. Anatomy. 2. Plant anatomy. I. Marshall Cavendish Corporation. II.
Title.

  QL805.A55 2006
  571.3--dc22

                                        2005053193

Printed in China
09 08 07 06 1 2 3 4 5

MARSHALL CAVENDISH
Editor: Joyce Tavolacci
Editorial Director: Paul Bernabeo
Production Manager: Mike Esposito

THE BROWN REFERENCE GROUP PLC
Project Editor: Tim Harris
Deputy Editor: Paul Thompson
Subeditors: Jolyon Goddard, Amy-Jane Beer, Susan Watts
Designers: Bob Burroughs, Stefan Morris
Picture Researchers: Susy Forbes, Laila Torsun
Indexer: Kay Ollerenshaw
Illustrators: The Art Agency, Mick Loates, Michael Woods
Managing Editor: Bridget Giles

Contents

Crocodile

ORDER: **Crocodylia** FAMILY: **Crocodylidae** GENUS: *Crocodylus*

Fourteen species of true crocodiles live in the tropics, mostly in rivers and large lakes. There are a further nine species of closely related dwarf crocodiles, alligators, caimans, and gharials. Together with the true crocodiles, they are called crocodilians. All live in tropical climates except for the American alligator, which lives as far north as Arkansas and North Carolina, and the critically endangered Chinese alligator of the Chang River.

Anatomy and taxonomy

Crocodiles are reptiles. Other living reptiles include lizards, amphisbaenians (worm lizards), snakes, turtles, and tuataras.

● **Animals** All animals are multicellular and rely on other organisms for food. They differ from other multicellular life-forms in their ability to move around (generally using muscles) and their rapid responses to stimuli.

● **Chordates** At some stage in its life, a chordate has a stiff, dorsal (back) supporting rod called a notochord.

● **Vertebrates** The notochord of vertebrates develops into a backbone (also called spine or vertebral column), which is made up of connected bones called vertebrae. The vertebrate muscular system that moves the head, trunk, and limbs consists primarily of muscles that are bilaterally symmetrical around the skeletal axis. In other words, the muscles on one side of the backbone are the mirror image of those that occur on the other side.

● **Reptiles** Reptiles are a group of four-legged vertebrates, although some have lost their legs during the course of evolution. They have thick, horny, waterproof skin that is usually divided into plates called scales. Most reptiles lay eggs with waterproof shells. Some female reptiles retain their eggs inside the body until the eggs are ready to hatch. A few snakes and lizards are viviparous—the mother and the embryo exchange nutrients and waste products. Reptiles are not endothermic (warm-blooded) animals like birds and mammals, but they do have some control over their body temperature. Reptiles can move into hot places when they need to warm up and into cooler ones

▶ *This family tree shows the major living crocodilian groups. The American crocodile is a true crocodile of the genus* Crocodylus.

Animals
KINGDOM Animalia

Chordates
PHYLUM Chordata

Vertebrates
SUBPHYLUM Vertebrata

Reptiles
CLASS Reptilia

Crocodilians
ORDER Crocodylia

Gharials
FAMILY Gavialidae

Crocodiles
FAMILY Crocodylidae

Alligators, caimans, and dwarf caimans
FAMILY Alligatoridae

Dwarf crocodile
GENUS *Osteolaemus*

True crocodiles
GENUS *Crocodylus*

when they need to lose heat. A reptile that has been lying in bright sunshine for a time may be warmer than a similarly placed bird or mammal.

● **Crocodilians** These large reptiles spend a lot of time in water, usually in rivers or large lakes. Crocodilians have a long body with a powerful tail. They are covered with large scales, underneath which lie bony plates called osteoderms, so these animals are very well-armored. Their jaws are long and bear pointed teeth. All crocodilians lay eggs. Biologists place them in a group called the order Crocodylia. There are four main types of crocodilians.

● **Alligators, caimans, and dwarf caimans** All the teeth of the lower jaw of these crocodilians fit into a row of pits in the upper jaw, so none can be seen when the mouth is closed. They also have relatively broad snouts. There are eight species.

● **Gharials** This family contains just one species. This is the gharial of northern India and Pakistan. It has a very long, narrow snout lined with long, pointed teeth, and feeds almost exclusively on fish.

● **African dwarf crocodile** This species lives in West Africa. It is comparatively small, and never grows to more than 6.5 feet (2 m) in length.

● **True crocodiles** These have an enlarged fourth tooth along each side of the lower jaw. Each of these teeth fits into a notch in the outer side of the upper jaw, so this tooth can still be seen when the jaws are fully closed. There are 14 species; the largest is the estuarine or

▲ *A crocodile basks on a riverbank. Basking is important for reptiles. It warms the body to a temperature at which activities such as movement and digestion can take place.*

saltwater crocodile of northern Australia and Southeast Asia, which can reach 20 feet (6 m) or more in length and weigh more than 1.5 tons.

● **American crocodile** This is one of the larger types of crocodiles, with males reaching 16.5 feet (5 m) long. They are armored with scales, but less so than other species. There is a distinctive swelling in front of each eye that occurs in all life stages except the hatchling stage. Juvenile American crocodiles have bands on the body and tail; adults are a uniform olive-brown color.

FEATURED SYSTEMS

EXTERNAL ANATOMY Crocodiles are large reptiles with four legs and thick, heavily armored skin. They have a pointed snout, and the jaws are lined with long, conical teeth. *See pages 150–152.*

SKELETAL SYSTEM Almost all the bones of crocodiles are massive. The bones of the skull are fused to form a solid protective cover for the brain and jaw muscles. *See pages 153–155.*

MUSCULAR SYSTEM Three sets of muscles are especially powerful, those of the jaws, the trunk, and the tail. *See pages 156–157.*

NERVOUS SYSTEM A crocodile's brain does not look very different from the brains of other reptiles, but crocodiles are more intelligent than most. They use a combination of sight, smell, and unique ripple-detecting bumps on their jaws to find food. *See page 158.*

CIRCULATORY AND RESPIRATORY SYSTEMS The heart of a crocodile has four chambers, although separation of oxygenated (oxygen-rich) blood from the lungs and oxygen-depleted blood from the body tissues is not complete. Crocodiles have two systemic arteries. They can breathe when their mouth is open underwater because air from the nostrils is separated from the mouth by a secondary palate. *See pages 159–160.*

DIGESTIVE AND EXCRETORY SYSTEMS The large tongue can be used to block the gullet while a crocodile is feeding underwater. The stomach is in two parts: the muscular gizzard and the glandular pyloric section. The last part of the alimentary canal is called the cloaca. The two kidneys are long, thin, and many-lobed. *See page 161.*

REPRODUCTIVE SYSTEM Male crocodiles have a single penis, which is housed in the cloaca. Adult female crocodiles lay eggs. *See pages 162–163.*

External anatomy

CONNECTIONS

COMPARE the hide of a crocodile with the skin of a **GREEN ANACONDA**.

COMPARE the teeth and jaws of a crocodile with those of other ambush predators, such as a **LION**.

Crocodiles are long, relatively thin animals with a long tail but comparatively short legs. Crocodiles can move fast on land, raising their body off the ground so the belly is supported only by the four legs; most lizards walk with the legs sprawled out sideways, so their belly is dragged along the ground as they move. Moving in this way uses a lot of energy, so crocodiles moving rapidly on land tire quickly and can keep up a fast speed only over short distances.

In water, crocodiles tuck their legs into the side of the body. A crocodile's long, thin shape and relatively pointed snout make it streamlined. Propulsion comes mostly from side-to-side movements of the powerful tail. Crocodiles can swim swiftly, silently, and for long periods. Crocodiles move in a different way when they are swimming slowly. Then they paddle with their webbed feet. The webbing helps improve the efficiency with which the feet can push against the water. Webbed feet are very common in mammals, birds, reptiles, and amphibians that spend a lot of time swimming in water.

Long, narrow snouts

The snouts of crocodilians are long and thin, although the width of the snout varies between species. Gharials, false gharials, and some caimans have a very thin snout. This is an adaptation for feeding on fish, which are captured by slashing the snout sideways with the jaws open. Other crocodilians feed on a wider variety of prey. Their snout is wider to incorporate the thick bones and powerful muscles that are needed to subdue other types of animals, with alligators having the widest snouts of all. Crocodilian eyes and nostrils are always raised above the level of the rest of the snout. This enables crocodiles to float with just the eyes and nostrils above the surface, so they can see and smell their prey and carry on breathing while remaining largely hidden.

The eyes of crocodiles have vertical pupils during the daytime. These widen during the evening or at night, allowing as much of the sparse light as possible into the eye. Crocodiles can therefore hunt at these times, when prey animals are often more active than during the heat of the day. Crocodile eyes are covered

▼ **American crocodile**
This large crocodilian lives from southern Florida through the Caribbean to northern South America.

The tough, scaly **hide** has scales, each of which overlies a bony plate.

Keeled scales along the tail aid stability as the animal swims.

The **tail** is used to propel the crocodile through the water.

Feet are important for slow sculling through water.

The **eyes** and nostrils are both dorsally located. This allows the animal to be almost completely submerged while remaining able to see and breathe above the water.

The **jaws** are the most powerful in the animal kingdom. The teeth are used to hold prey and to rip chunks of flesh from the corpse.

3.5 feet (1 m)

20 feet (6 m)

by a transparent membrane. This is important when the head is underwater, since it protects the delicate surface of the eyeball. The openings of the nostrils and the ears are also protected. Complicated valves close when the animal is submerged. Each ear has a small flap of soft skin. This can be moved by muscles to block the ear opening so water cannot enter and damage the eardrum.

Fearsome teeth

All crocodilians have rows of large, conical teeth along the edges of the upper and lower jaws. The teeth are not all the same size; the first and fourth pairs in the lower jaw are usually longer than the rest. The first pair projects in front of the snout. In crocodiles, the fourth pair projects outward. The teeth fit into a groove on the outside of the upper jaw, so they can still be seen when the jaws are closed. Other crocodilians lack this groove. Instead, the fourth pair of teeth slots into pits in the upper jaw; the teeth cannot be seen when the jaws are shut. Teeth in the upper and lower jaws in all crocodilians alternate, so they intermesh when the jaws are closed.

Crocodile teeth are regularly replaced. Each tooth usually has the bud of a new one growing under its root. As this bud grows into a new tooth, it forces the old one to detach and fall out. Crocodiles replace teeth throughout their lives, but the rate of replacement slows in older animals.

The gharial's snout

The tip of the snout in adult male gharials is expanded into a knob (right). This is caused by the enlargement of the bone at the front of the upper jaw, the premaxilla. The gharial's knob is called a bulla. It is hollow and has a complicated structure that involves parts of several other bones, such as the palatine. The bulla can be used to produce sounds that are involved in courtship.

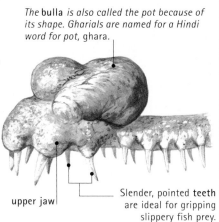

The bulla *is also called the pot because of its shape. Gharials are named for a Hindi word for pot, ghara.*

upper jaw

Slender, pointed **teeth** are ideal for gripping slippery fish prey.

Armor-plated reptiles

Reptiles have evolved a number of forms of armor. The scales have become thick in some lizards, such as the alligator lizards of the southwestern United States. Bony plates lie beneath each scale. These bony plates are called osteoderms. Crocodilians also have osteoderms, mostly beneath the scales on the back and sides. The scales of crocodilians are hard and horny, but the spaces between them are made of softer skin. Apart from their eyes, this soft skin is the animals' sole point of vulnerability. It is the only place where

COMPARATIVE ANATOMY

Separating crocs and gators

When a crocodile closes its mouth, most of the teeth on the upper jaw can still be seen, since they lie outside the edge of the lower jaw. Most of the teeth of the lower jaw cannot be seen, as they fit into holes in the bones of the upper jaw. The fourth tooth along from the front is larger than the rest, however, and projects outward. It fits snugly into a groove in the side of the upper jaw, but remains visible. When a crocodile or alligator becomes very old, the bone around the holes in the upper jaw and the gums wears away, so this distinction is not so clear-cut. However, a further difference between crocodiles and alligators helps biologists tell the two groups apart. From above, the snout is long and thin in crocodiles but it is more rounded in alligators.

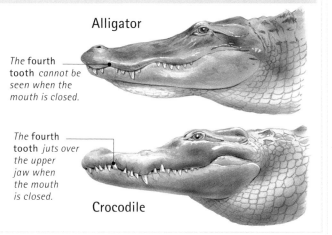

Alligator

The **fourth tooth** *cannot be seen when the mouth is closed.*

The **fourth tooth** *juts over the upper jaw when the mouth is closed.*

Crocodile

Communicating by smell

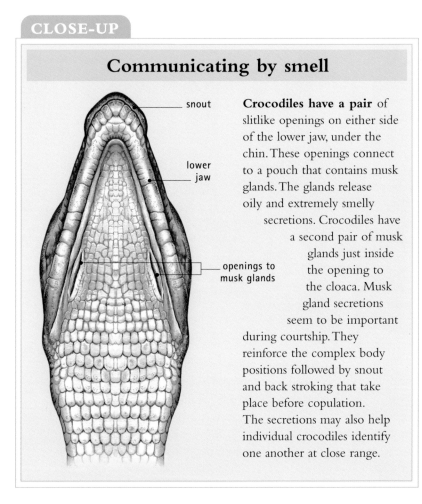

snout

lower jaw

openings to musk glands

Crocodiles have a pair of slitlike openings on either side of the lower jaw, under the chin. These openings connect to a pouch that contains musk glands. The glands release oily and extremely smelly secretions. Crocodiles have a second pair of musk glands just inside the opening to the cloaca. Musk gland secretions seem to be important during courtship. They reinforce the complex body positions followed by snout and back stroking that take place before copulation. The secretions may also help individual crocodiles identify one another at close range.

external parasites, such as ticks or leeches, are able to gain a hold. In turtles, the scales have fused to form a solid shell supported by bones that mostly evolved from expanded ribs.

Scale variations

Not all crocodile scales are the same size. Those on the head are usually smaller and softer than elsewhere. Scales on the back may have longitudinal ridges. These scales are described as "keeled." Keeled scales line up in rows that run along the length of the animal.

Crocodiles usually have five rows of keeled scales in the middle of the back. At the top of the front end of the tail, two rows of keels are especially large and are often pointed. They look almost like teeth. These two rows of keeled scales fuse about halfway along the tail, so there is then just one row running along the top in the center. Keeled scales mainly serve for protection, but they may help stabilize the animal when it is swimming quickly, like the keel of a boat.

▼ The white patch on the tongue of this American crocodile is formed by the lingual salt glands. Absent in alligators, these glands secrete salt and allow crocodiles to live in a wider range of habitats.

Skeletal system

All parts of the skeleton of crocodiles are made of strong, heavy bones. The bones of the skull are fused to form a massive protective box, which has a far more solid-looking appearance than the skulls of most lizards and snakes. The bones that lie underneath the scales on the top and sides of the skull are the osteoderms. They may be joined to the skull bones by connective tissue. This arrangement gives the bones of the surface of the skull a pitted, spongy appearance, which becomes more pronounced as a crocodile grows old.

Teeth and jaws

The teeth in the upper jaws are fixed to two bones, the premaxilla and the maxilla. They are joined together firmly. On each side of the head the premaxilla usually has five teeth and the maxilla 14. The lower jaw has 14 or 15 teeth on each side. They are all fixed to a single bone, the dentary (also called the mandible). The lower jaw also contains a number of other bones, the most important of which are the surangular and the angular bones. The individual bones of the lower jaw are so firmly fused to one another, however, that they look and act like one bone.

The main muscles that close the jaws run from the lower jaw to the outside surface of the bones of the braincase. They occupy the space beneath the jugal bone, which helps protect them. The articulation (place where movable bones join) of the lower and upper jaws is between the articular (the rear bone of the lower jaw) and the quadrate bones. Muscles attached to the part of the articular bone that lies behind the jaw hinge (the retro-articular process) run forward to attach to the braincase; these muscles open the jaws.

The teeth of crocodiles are subjected to a lot of wear, and eventually they must be replaced. Each tooth usually has the bud of a new one growing under its root. As this bud grows into a new tooth, it forces the old one to detach. Crocodiles go on replacing teeth throughout life, but experts think that the rate at which this happens slows as an individual gets older.

Crocodile jaws

Crocodilians that feed exclusively on fish have long, thin jaws, and most of the teeth are approximately the same size. Fish feeders also have more teeth; the gharial has 28 teeth on each side of the upper jaw and about 25 on each side of the lower jaw.

Crocodilians that take a wider range of prey have wider snouts. The American alligator has a wide snout because it needs strength to grab and overcome large, powerful prey, such as mammals and other alligators. American crocodiles have jaws that are intermediate between these two jaw types. They represent an evolutionary compromise: American crocodiles need powerful jaw muscles and a solid skull to capture large mammal prey, but they also need to be able to catch fish when other types of prey are scarce.

Alligator

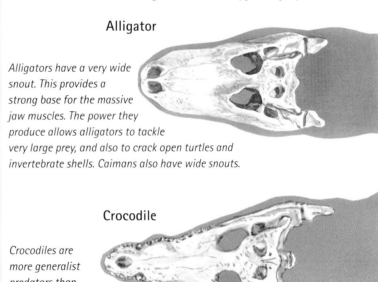

Alligators have a very wide snout. This provides a strong base for the massive jaw muscles. The power they produce allows alligators to tackle very large prey, and also to crack open turtles and invertebrate shells. Caimans also have wide snouts.

Crocodile

Crocodiles are more generalist predators than alligators. They can tackle large animals, but their narrower snouts allow them to take fish and other smaller prey when food is scarce.

Gharial

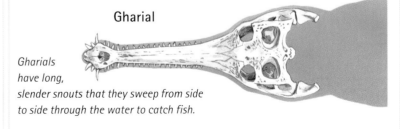

Gharials have long, slender snouts that they sweep from side to side through the water to catch fish.

▲ *A male gharial exposes his teeth. Compare these with the teeth of an American crocodile.*

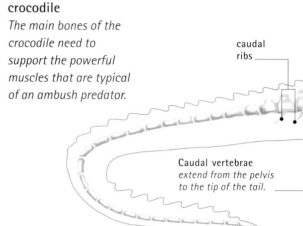

▶ **American crocodile**
The main bones of the crocodile need to support the powerful muscles that are typical of an ambush predator.

caudal ribs

Caudal vertebrae *extend from the pelvis to the tip of the tail.*

Vertebrae and ribs

In crocodiles, the individual bones of the backbone, the vertebrae, are massive. That is because they have to support the weight of the heavy body when it is raised off the ground. There are usually 26 vertebrae between head and pelvic girdle, and more than 35 vertebrae in the tail. The first two ribs are movable but the next five are fixed to the sternum. The two sacral vertebrae are fused to

give extra strength to the pelvis. Some of the bones at the base of the tail have projections at the side called caudal ribs.

The shoulder girdle supports the front legs. The two coracoids attach to the backbone; the two scapulae are mainly for the attachment of the muscles that operate the forelimbs. The scapulae are flat, rectangular bones. The pelvic girdle is more complicated and has three main bones on each side: Each ileum attaches the girdle to the backbone. Each ischium joins its opposite number at the lower part of the body, and each pubis projects backward.

The pelvic girdle of a crocodile does not look much like that of a mammal—it is much more like the hip girdle of a bird. Crocodiles and their fossil relatives are unusual in that the pubis does not play a part in forming the

EVOLUTION

The biggest crocodile ever

The biggest crocodile known to science is called *Sarcosuchus imperator* (below). It lived 110 million years ago in West Africa. Adults could grow to nearly 40 feet (13 m) in length and weighed more than 8 tons. The skull alone was almost 6 feet (1.8 m) long. Biologists can use

Sarcosuchus's bone dimensions to figure out how this mighty beast may have lived. *Sarcosuchus* had robust teeth and a wide snout, suggesting that it was no fish eater. It seems that *Sarcosuchus* ambushed small- to medium-sized dinosaurs as they drank or tried to cross waterways.

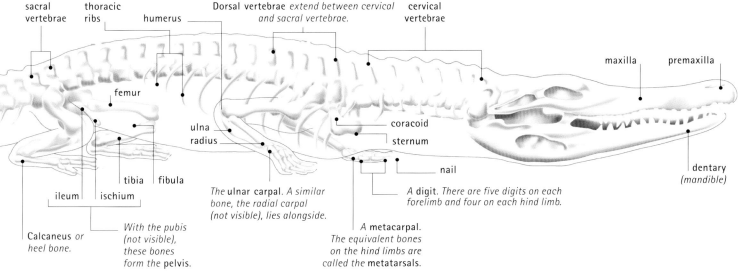

sacral vertebrae

thoracic ribs

humerus

Dorsal vertebrae *extend between cervical and sacral vertebrae.*

cervical vertebrae

maxilla

premaxilla

femur

ulna
radius

coracoid
sternum

nail

dentary
(mandible)

tibia | fibula

ileum | ischium

The ulnar carpal. *A similar bone, the* radial carpal *(not visible), lies alongside.*

A digit. *There are five digits on each forelimb and four on each hind limb.*

Calcaneus *or heel bone.*

With the pubis *(not visible), these bones form the* pelvis.

A metacarpal. *The equivalent bones on the hind limbs are called the* metatarsals.

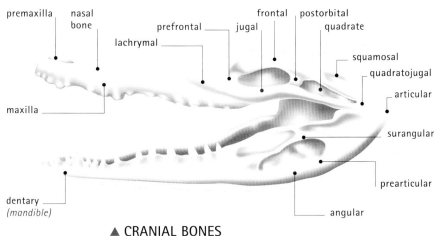

premaxilla

nasal bone

prefrontal

lachrymal

frontal

jugal

postorbital

quadrate

squamosal

quadratojugal

articular

maxilla

surangular

dentary
(mandible)

prearticular

angular

▲ CRANIAL BONES
American crocodile
The bones of the skull are very robust to provide anchorage for the supremely powerful jaw musculature.

"cup" into which the top of the femur fits to make a ball-and-socket joint. Scientists do not fully understand the reason for this.

Bones of the leg

The leg bones of crocodiles must be strong, since they need to bear the weight of the body on land. The main bones of the foreleg comprise the humerus followed by the radius and the ulna; in the hind leg the femur is followed by the tibia and the fibula. All of these bones are short and stout. By contrast, the metatarsals, metacarpals, and bones of the digits are relatively long and thin. This is because they make up the feet, which, as well as supporting the weight of the animal, form the paddles with which crocodiles move when swimming slowly through the water.

EVOLUTION

Ancient crocodiles

Crocodiles appeared about 215 million years ago, in the middle of the Triassic period. Like modern crocodiles, these ancient crocs lived partly on land and partly in the water. Their snouts were much shorter than those of modern crocodiles, and they measured only about 3 feet (1 m) long. Many kinds of crocodiles evolved from these small ancestors. Some had longer legs than modern crocodiles, with hooflike feet

for running on land; some were even bipedal. Others had elaborate armor formed from hardened scales. Some ancient crocodiles had snouts that were even longer and thinner than those of modern gharials. Some of these fish-eating crocodiles lived in the sea; they never left the water except to lay their eggs.

▼ **Desmatosuchus** *was an ancient relative of crocodiles. It was an aetosaur, or "eagle lizard," a group named for their winglike spikes. Unlike true crocodiles, Desmatosuchus was a plant eater with peglike teeth.*

Muscular system

Crocodiles have extremely powerful jaws. These, together with their large pointed teeth, makes them fearsome predators. The most important jaw muscle is called the intramandibularis muscle. This attaches to the braincase and runs to the lower jaw. It can contract rapidly and is the main reason that crocodiles can close their jaws with a powerful, fast snap. Longer muscles called pterygoids run from the back of the articular bone to the upper jaw. This adds to the force exerted by the intramandibularis muscle.

The jaws are opened partly by a muscle that runs from the front of the articular bone to the back of the skull. This is called the mandibular depressor muscle. When this muscle contracts, the back of the lower jaw is pulled upward and forward. Because the hinge of the jaw is in front of this, the front part of the lower jaw moves downward, causing the mouth to gape.

The mandibular depressor muscle is much smaller and less powerful than the muscles that close the jaws, but it is powerful enough to enable a crocodile to open its mouth very wide so large chunks of food can be swallowed. The size of a crocodile mouthful is limited not by how far apart the animal can open its jaws but by the width of its gullet. This is because crocodiles do not chew their food. However, they do break it into

▼ **American crocodile**
The major muscles and muscle groups of an American crocodile.

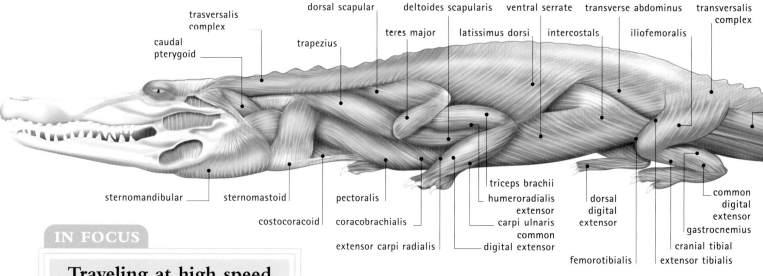

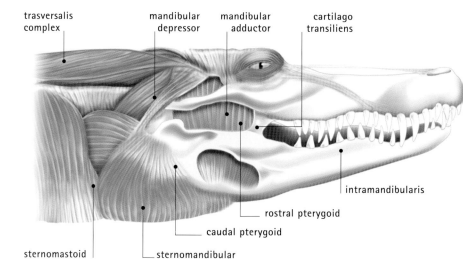

IN FOCUS

Traveling at high speed

Smaller crocodiles sometimes gallop for high-speed escape. Johnston's crocodile from Australia is a frequent galloper. Both front legs are raised off the ground while the hind legs propel the animal forward. The front legs then fall to the ground, providing further propulsion as the hind legs are raised. Sometimes a galloping crocodile moves so quickly that there are periods when all four feet are off the ground at the same time— just like a galloping horse. Galloping is fast, but it uses a huge amount of energy, so a Johnston's crocodile cannot gallop for long.

Changing gaits

Like other land vertebrates, crocodiles change the way they move depending on how fast they need to travel. These movement strategies are called gaits. At low speeds, crocodiles crawl. The body flexes from side to side; one front leg and the opposite back leg move forward while the other front leg and opposite back leg move back relative to the body. The feet of the forward-moving legs are raised, while those moving backward remain on the ground. The crocodile moves forward as a result, with the belly dragging along the ground. This gait is sometimes called the "belly slide." At higher speeds, crocodiles raise the belly off the ground, and the flexions of the body and the movements of the legs become more pronounced. This is called a "high walk;" it allows faster movement but uses much more energy.

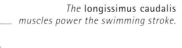

The **longissimus caudalis** *muscles power the swimming stroke.*

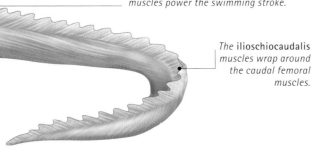

The **ilioschiocaudalis** *muscles wrap around the caudal femoral muscles.*

▲ *A Nile crocodile attacks a wildebeest. Crocodilians have jaw muscles of astonishing power. An alligator can bite down with a force of up to 3,000 pounds (1,360 kg), by far the strongest bite in the animal kingdom. By comparison, sharks bite with a force of 330 pounds (150 kg). Humans can muster only 175 pounds (80 kg), and dogs only 125 pounds (57 kg).*

bite-sized pieces, by shaking it and by rotating their body in the water until the piece of prey that is grabbed twists and tears free.

Trunk, limbs, and tail

The trunk muscles need to be powerful since a crocodile uses them to bend its body into an S-shape. By tucking its legs into its sides, the crocodile can then swim through the water like a fish. These trunk muscles are collectively called the iliocostal muscles. They form large blocks running down the side of the body; each iliocostal muscle corresponds to a rib, although there is no direct connection.

The muscles running along the backbone also contribute to the bending required for movement. The muscles of the limbs of crocodiles are very complicated but they do not differ significantly from the leg muscles of similar quadrupedal vertebrates (those that walk on four legs).

Crocodiles have very powerful tail muscles. The longissimus caudalis muscles contract alternately to lash the tail from side to side during swimming. Crocodiles may also use their tails in defense. A crocodile usually has between 35 and 45 individual bones, or vertebrae, that form the backbone of the tail. Most of the individual muscles run from one vertebra to the next. Each vertebra has projections at the top, bottom, and sides that act as attachments for these muscles.

Nervous system

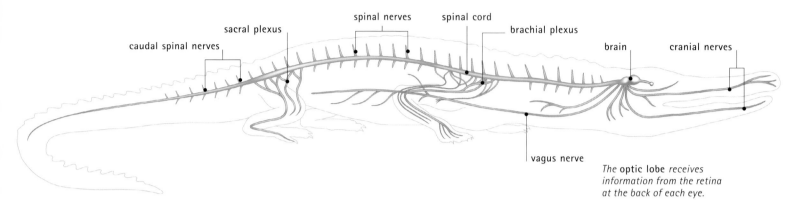

caudal spinal nerves · sacral plexus · spinal nerves · spinal cord · brachial plexus · brain · cranial nerves · vagus nerve

The optic lobe *receives information from the retina at the back of each eye.*

cerebral hemisphere · diencephalon · cerebellum

olfactory tract

The auricle *is important for hearing.*

second cranial nerve

infundibulum

The olfactory bulb *interprets information relating to smell and taste.*

third cranial nerve

The pituitary gland *is crucial for endocrine (hormonal) function.*

medulla oblongata

12 pairs of cranial nerves govern movement in the head.

▲ American crocodile
The central nervous system and some elements of the peripheral nervous system are shown.

▶ BRAIN American crocodile
This view shows only one of each olfactory bulb and optic lobe.

The various parts of a crocodile's brain occur in sequence, as in the brains of other lower vertebrates. Most of the main parts of the brain are paired. The first part comprises the two olfactory bulbs. They lie toward the front of the snout, near to the nostrils from which they receive and interpret sensory information. Crocodiles have a good sense of smell; this helps them find putrefying carrion on land or food that they have lodged underwater.

Intelligent reptiles

Reptiles are smarter than amphibians but much less intelligent than most mammals. Most scientists consider crocodiles to be the most intelligent reptiles. Crocodiles display highly developed social behavior and, unlike lizards, snakes, or turtles, they communicate with one another using sounds. These include grunts and roars, which humans can hear, and low-frequency infrasonic sounds that travel underwater; these are beyond the auditory ability of humans. Intelligent behavior in vertebrates is related to the size and surface area of the cerebral hemispheres. These are the biggest parts of the crocodile brain.

Two smaller lobes—the optic lobes—lie immediately behind the cerebral hemispheres. The optic lobes register nerve impulses from the optic nerves, which bring information from the retina at the back of each eye. Unlike amphibians and reptiles such as lizards, crocodiles do not have a light-sensitive section in the brain that controls biological rhythms.

Controlling movement

Behind the optic lobes lies the unpaired cerebellum. It is the region of the brain where the various senses and muscle contractions needed to permit movement are coordinated. Finally, the hindbrain, or medulla oblongata, runs backward to merge into the spinal cord. Nerve impulses from the ears are received and interpreted near the front of the hindbrain, but there is not a distinct lobe devoted to hearing.

CLOSE-UP

Detecting ripples

Crocodiles and alligators have a dusting of bumps over their jaws. These are packed with nerve endings, so they are clearly sensory in function, but the nature of their importance was not discovered until 2002. The bumps are ripple detectors of great sensitivity. The bumps allow a submerged crocodile to pinpoint splashes caused by an animal drinking or falling into the water. Crocs may rely on this sensitivity more than eyesight or smell to find prey.

Circulatory and respiratory systems

The most important part of the circulatory system is the heart, which pumps blood around the body. Crocodiles have a heart that is almost—but not quite—divided into four parts. Oxygen-rich blood from the lungs enters the left atrium through the pulmonary veins. From there it is pumped into the left ventricle. Crocodiles are the only reptiles in which the left ventricle is separated from the right ventricle. The right systemic artery arises from the left ventricle. It carries oxygenated blood to the body. The blood passes through a network of increasingly small vessels until, in the capillaries, it gives up its oxygen to tissues. Once it has lost its oxygen, blood is carried by veins back to the heart. It enters the right atrium and is then pumped into the right ventricle. Two major arteries lead from the right ventricle. One is the pulmonary artery, which carries blood to the lungs, where it picks up more oxygen. The other is the left systemic artery, which runs around the heart on the left hand side of the animal before joining with the right systemic artery.

An important little hole

There is a small hole between the right and left systemic arteries. Near the heart, these arteries run side by side. The hole is called the foramen of Panizza. Studies using radioactive tracers and other techniques have shown that much of the blood in the left systemic artery comes not from the right ventricle, but from the left ventricle, having passed through

CLOSE-UP

Unique heart valves

The valves at the top of the right ventricle in Australian estuarine crocodiles are unique. Flaps of connective tissue on their surface can intermesh; they are called teeth owing to their resemblance to the teeth of a cogwheel. When the crocodile is relaxed the teeth mesh together; most of the blood pumped out by the right ventricle goes to the body. However, when the animal is under stress the hormone epinephrine causes the teeth to contract. This allows more blood to be pumped to the lungs, where it picks up oxygen. The oxygenated blood is then pumped to the tissues, enabling the crocs to sustain more vigorous activity.

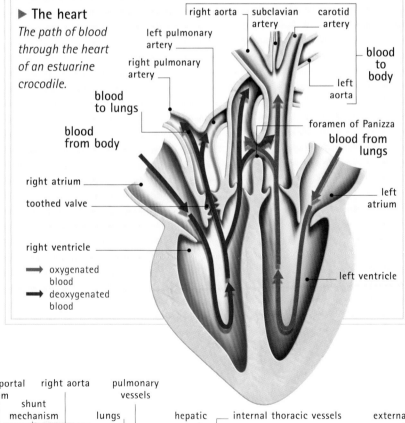

▶ **The heart**
The path of blood through the heart of an estuarine crocodile.

blood to lungs

blood from body

right aorta · subclavian artery · carotid artery

left pulmonary artery

right pulmonary artery

blood to body

left aorta

foramen of Panizza

blood from lungs

right atrium

toothed valve

right ventricle

left atrium

left ventricle

➡ oxygenated blood
➡ deoxygenated blood

▼ **American crocodile**
This diagram shows how blood flows through a crocodile's body. Sections labeled "vessels" include both veins and arteries.

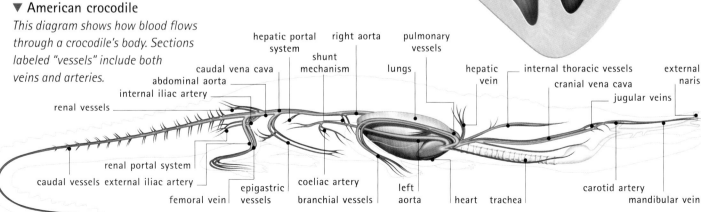

renal vessels

abdominal aorta
internal iliac artery

caudal vena cava

hepatic portal system

shunt mechanism

right aorta

pulmonary vessels

lungs

hepatic vein

internal thoracic vessels

cranial vena cava

jugular veins

external naris

renal portal system

caudal vessels · external iliac artery

femoral vein

epigastric vessels

coeliac artery

branchial vessels

left aorta

heart

trachea

carotid artery

mandibular vein

Breathing underwater

The lungs of crocodiles are well developed, with alveoli more tightly packed than in other reptiles. Alveoli are bundles of tissue through which gases are exchanged. A muscle called the diaphragmaticus powers breathing. This links the pelvis to the liver; when this muscle pulls on the liver, the lungs expand, and the crocodile breathes in. The abdominal muscles pull the liver forward to allow exhalation.

Crocodiles have a bony secondary palate that they use to separate the air passages from the mouth. Air is sandwiched between the secondary palate and the pterygoid bone. The air passage opens at the back of the throat. The closure of this secondary palate allows a crocodile to breathe while lying in the water with its nostrils above the surface and its mouth open. The secondary palate prevents water from entering the throat even if the crocodile is thrashing about in the water while subduing prey.

▲ *The location of the nostrils on top of the snout allows this spectacled caiman to breathe while almost completely submerged.*

the foramen. Experiments using pressure recorders inserted into the heart and arteries have shown that a crocodile can control the amount of blood being pumped to the lungs or to the body; when the animal dives less blood goes to the lungs. Crocodiles can also use this flow mechanism, which is called a shunt, to help regulate body temperature. If a crocodile gets too hot because it has been basking for too long, more blood can be sent to the skin to help carry excess heat away.

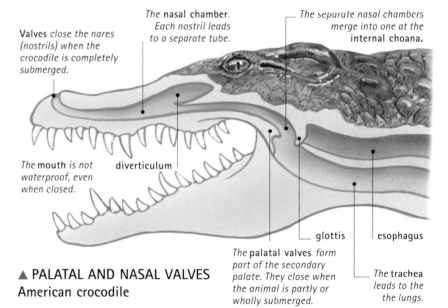

Valves close the nares (nostrils) when the crocodile is completely submerged.

The nasal chamber. Each nostril leads to a separate tube.

The separate nasal chambers merge into one at the internal choana.

The mouth is not waterproof, even when closed.

diverticulum

glottis   esophagus

The palatal valves form part of the secondary palate. They close when the animal is partly or wholly submerged.

The trachea leads to the the lungs.

▲ PALATAL AND NASAL VALVES
American crocodile

COMPARATIVE ANATOMY

Crocodile arteries and portal systems

In addition to the foramen of Panizza with its toothed valve, there are two other main ways in which the crocodile circulatory system differs from that of birds or mammals. The first is that there are two systemic arteries leaving the crocodile's heart; birds and mammals have only one. The second is that crocodiles have two portal systems. Portal systems are sections of the circulatory system in which blood runs through veins connecting different sets of capillaries together. All vertebrates have a hepatic portal system, which takes nutrient-rich blood from the small intestine to the liver. Fish, amphibians, and reptiles such as crocodiles also have a renal portal system. This carries blood from tissues in the rear part of the body to the kidneys. In birds and mammals the renal portal system is replaced by renal arteries.

Digestive and excretory systems

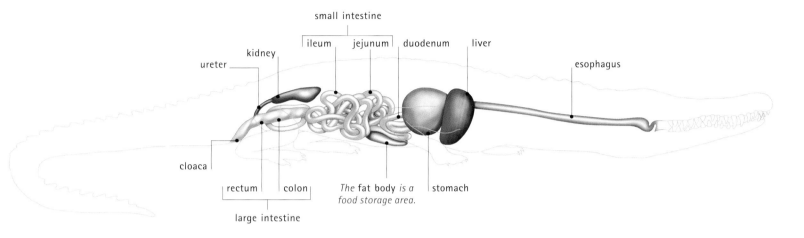

small intestine

ileum | jejunum | duodenum | liver

kidney

ureter

esophagus

cloaca

rectum | colon

The fat body *is a food storage area.* | stomach

large intestine

Crocodiles do not chew their food but instead swallow it whole. A crocodile has a large tongue, but this cannot be protruded like the tongue of a lizard or a snake. The tongue contains mucus-producing glands. The mucus helps the animal swallow large items of prey; the biggest crocs can swallow a pig or a human whole. The back end of the tongue can be used to block the top of the esophagus, aided by a flap of muscle at the back of the palate (the top of the mouth). This enables the animal to grab its prey underwater without flooding the digestive system.

To the stomach and beyond

The stomach is comparatively small. If a large animal has been swallowed as prey, it may remain in the gullet (esophagus) until there is room in the stomach. Crocodile stomachs are similar to those of birds in that they are divided into two distinct parts. The first is the gizzard, which is muscular and has a horny lining. Food is ground up in the gizzard, which often contains stones called gastroliths that help with the process. The second, pyloric part of the stomach does not have a horny lining. Digestive glands in its walls release their contents to break down the food.

The first part of the intestine following the stomach is the duodenum. This is followed by the small intestine; the demarcation between these two regions is the point at which the bile ducts from the liver and the pancreatic ducts enter. Unlike many vertebrates, crocodiles do not have a dead-end part of the intestine, or cecum. This is because they do not feed on plants, which are harder to digest than animal material. Vertebrates that eat plants need a part of the intestine where the resistant cell walls can be attacked by single-celled organisms; this often occurs in a cecum.

The next parts of the digestive system are the large intestine and the hindgut, or rectum. The rectum opens into the baglike cloaca. The cloaca has two glands attached to it that produce an oily, smelly secretion. These are called musk glands—there are also musk glands under the chin. Crocodiles remove waste from the blood in the kidneys. They are long, thin, and many-lobed. The kidneys open through two tubes called ureters into the cloaca, through which wastes are released.

▲ American crocodile

Note that both solid and liquid waste leave through the cloaca. The cloaca also serves as the exit for eggs in females and sperm in males.

COMPARATIVE ANATOMY

Reptiles in salt water

If reptiles go into the sea or into salty water in estuaries, they absorb salt and also take it in when they swallow. They get rid of this salt in a number of different ways. Marine iguanas in the Galápagos Islands have salt-secreting glands in their nostrils. Marine turtles use modified tear glands near the eyes. Crocodiles secrete the salt from glands in their tongue. They do not swallow the salty secretions, as the top of the gullet is blocked off by a flap at the back of the mouth. Alligators do not have salt-secreting tongue glands, and so cannot go into salty water. This explains why alligators are less widely distributed than crocodiles.

Reproductive system

All crocodiles reproduce by laying eggs. These are produced in the two ovaries of the adult female. Crocodile ovaries are long and flattened. The eggs develop in the oviducts. When they are ready to be laid they pass into the cloaca, which doubles as the last part of the alimentary canal. Between 20 and 60 eggs are laid in a nest, which is usually a mound of dead and decaying vegetation that the female builds near a river. The eggs hatch after about 90 days of development in the warm nest. Juvenile American crocodiles are about 8 inches (40 cm) long when they hatch.

The cloaca

Sperm are produced in the male's testes, which are long and oval, and is carried to the cloaca along sperm ducts, the vasa deferentia. The cloaca of a crocodile has a complicated structure. A first chamber, or coprodaeum, is used to store waste material until it is ready to be evacuated as a fecal mass. There is a ring of muscle called a sphincter at the end of the coprodaeum. When this contracts, feces are prevented from passing into the crocodile's rectum. Immediately beyond the coprodaeum is a second baglike structure called the urodaeum. The two ureters carrying nitrogenous waste from the kidneys open into the urodaeum. In females, the oviducts also open into the urodaeum.

At the front of the urodaeum is a second, less muscular sphincter. The final chamber of the cloaca is called the proctodaeum. The male's penis is housed in the proctodaeum. The vasa deferentia open near the base of the penis. The opening of the crocodile cloaca is a slit that runs lengthwise on the underside of the body at the base of the tail. This opening is called the vent. The vent in lizards and snakes has a different structure; it runs from side to side.

Courtship and mating

Crocodiles and turtles have a single penis; other reptiles, like lizards and snakes, have two. Crocodile courtship is often elaborate. After

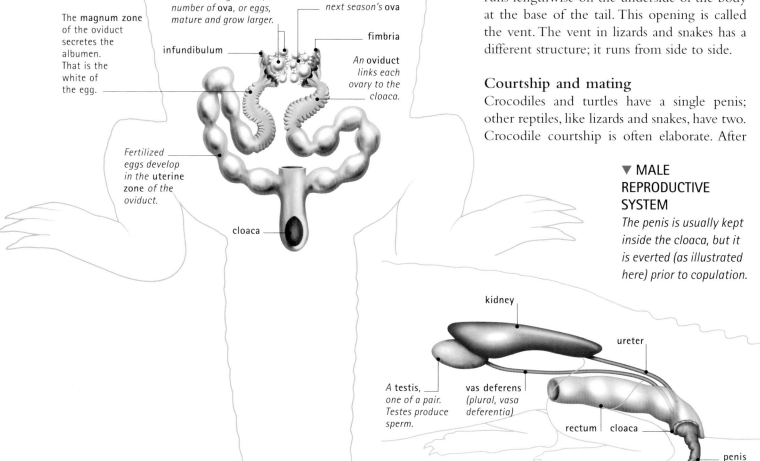

▼ EGGS AND OVIDUCTS
The female crocodile reproductive system. The fimbria are filaments that carry an ovum from the ovary into the oviduct.

The **magnum zone** of the oviduct secretes the albumen. That is the white of the egg.

*Each breeding season, a number of **ova**, or eggs, mature and grow larger.*

infundibulum

*next season's **ova***

fimbria

*An **oviduct** links each ovary to the cloaca.*

Fertilized eggs develop in the **uterine zone** of the oviduct.

cloaca

▼ MALE REPRODUCTIVE SYSTEM
The penis is usually kept inside the cloaca, but it is everted (as illustrated here) prior to copulation.

kidney

ureter

A testis, one of a pair. Testes produce sperm.

vas deferens (plural, vasa deferentia)

rectum cloaca

penis

1

A male and a female alligator attract each other by bellowing. The jaws are slapped onto the water.

2

Male and female touch snouts and make coughing sounds. They engage in a range of courtship rituals, including blowing spouts of water from the nostrils.

courtship, the male climbs onto the female's back. He bends the underside of his pelvic region and the base of his tail so that the opening of his cloaca lies next to the female's. His penis is then inserted into the female's cloaca. The penis has a groove, and sperm runs along it into the female. This differs from a mammal's penis, in which sperm runs along a tube called the urethra. Crocodile copulation may last for up to 15 minutes.

ROGER AVERY

3

Copulation begins as the male mounts the female. Sperm is discharged from the penis. It enters the cloaca of the female and will travel up the oviduct to fertilize the ova.

▲ COURTSHIP AND MATING
American alligator

FURTHER READING AND RESEARCH

Harris, T. (ed.). 2003. *Reptiles and Amphibians.* Marshall Cavendish: Tarrytown, NY.

CLOSE-UP

Crocodile eggs

The eggs of crocodiles take up to 13 weeks to hatch (right). During this period the embryo grows from a single cell into a juvenile crocodile. The sex of a young croc is not determined by genes from the parents as in many other animals. Instead, the warmth at which an egg develops determines the embryo's sex. The embryo gains nutrients from the yolk. The yolk is surrounded by a thin-walled membrane, the yolk sac. Part of the yolk sac enters the developing embryo through the underside of its body, and blood vessels transfer yolk to the intestine.

Crocodile eggs have three other membranous sacs. Two, the amnion and the chorion, surround the embryo and help protect it. The final egg membrane is the allantois, which grows from the intestine and occupies part of the space between the chorion and the amnion. The allantois allows toxic nitrogenous excretory wastes to be stored well away from the developing embryo.

The chorion and allantois played an important role in the evolution of land vertebrates. They allowed reptiles to reproduce on land, and later became modified to form part of the placenta in mammals.

A cross section through an alligator egg.

The **yolk** *nourishes the embryo.*

developing embryo

allantois

chorion

The **outer shell** *is waterproof but does allow gases such as oxygen to move in and out.*

amnion

The **albumen,** *or egg white.*

Diatom

KINGDOM: **Protista** PHYLUM: **Stramenopiles**
CLASS: **Bacillariophyta**

Diatoms are part of a large and diverse group of mostly single-celled life-forms called protists. Protists live in every watery environment on Earth, including inside humans. Some protists are parasites. Diatoms are among the most beautiful of protists. Each diatom has an exquisite glassy covering composed of two parts that fit together like the lid and base of a tiny box.

Anatomy and taxonomy

In the five-kingdom system commonly used to classify life on Earth, the kingdom Protista includes all the single-celled animal-like and plantlike organisms that do not qualify as plants, animals, fungi, or prokaryotes (bacteria).

● **Protists** Members of the Protista are often called protists, meaning "first beings." Far from being simple, protist cells are often far more complex than those of multicellular organisms. After all, they must perform all essential processes of life—there is no division of labor as in so-called "higher" organisms. Protists do not have separate organs devoted to tasks such as locomotion, obtaining food, self-defense, maintaining water balance, and reproduction; instead, every function is performed by just one cell. Protist cells range in size from about one-thousandth of an inch (2 microns) to 1 inch (2.5 cm) in the case of some foraminiferans. There are also many colonial species in which individual cells continue to function independently but live in close association with others.

Protists are a very diverse group, divided roughly into plantlike and animal-like organisms. Animal-like protists are called protozoans, literally "first animals." They include amoebas and many parasites of humans. The plantlike protists are algae. Most algae are single-celled and microscopic, but this group also includes the seaweeds.

● **Green algae** Green algae are members of the plantlike phylum Chlorophyta and are typical algae. They live in a range of aquatic and terrestrial environments, from ponds to rocks and tree bark. Green algae have chlorophyll (green pigments) in the chloroplasts, cellulose, and stores of starchy and oily materials. Their overall form is highly variable, including unicellular, colonial (in which single cells live together, forming a larger mass), or multicellular organisms (such as seaweeds).

▶ *This family tree shows some of the major living protist groups. Diatoms, green algae, dinoflagellates, radiolarians, and foraminiferans are plantlike. Amoebas and heliozoans are animal-like.*

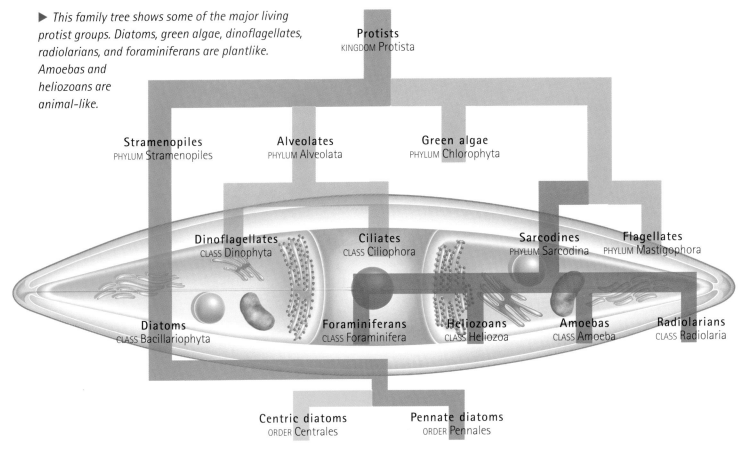

Protists
KINGDOM Protista

Stramenopiles
PHYLUM Stramenopiles

Alveolates
PHYLUM Alveolata

Green algae
PHYLUM Chlorophyta

Dinoflagellates
CLASS Dinophyta

Ciliates
CLASS Ciliophora

Sarcodines
PHYLUM Sarcodina

Flagellates
PHYLUM Mastigophora

Diatoms
CLASS Bacillariophyta

Foraminiferans
CLASS Foraminifera

Heliozoans
CLASS Heliozoa

Amoebas
CLASS Amoeba

Radiolarians
CLASS Radiolaria

Centric diatoms
ORDER Centrales

Pennate diatoms
ORDER Pennales

● **Dinoflagellates** Organisms in the division Dinophyta, or dinoflagellates, rival some diatoms in terms of microscopic grandeur and diversity. They are distinguished from other plantlike protists by a covering of thick cellulose and hemicellulose cell plates. There are two flagella; one (the transverse flagellum) encircles the cell in a groove called the girdle, and the second (the longitudinal flagellum) runs perpendicular to the girdle and often has a free end in the water. The second whips around to propel the cell. Most dinoflagellates are single-celled, normal reproduction is asexual, and energy reserves are stored as starch and fat.

● **Flagellates** Within the protist group Mastigophora, commonly known as flagellates, the distinction between plant and animal life is blurred. This large group is characterized by conspicuous flagella, cellulose cell walls, and starchy storage products. Most flagellates live in aquatic environments, and these vary from fresh water to the body fluids of larger organisms. The flagellates include some plantlike organisms that contain green-pigmented chloroplasts and obtain most of their nutrition by photosynthesis. However, others swim actively and consume organic material as food—for example, the mobile trypanosomes, which are parasitic in both plants and animals and cause a number of diseases in humans, such as sleeping sickness.

● **Ciliates** The Ciliophora, or ciliates, are one of the largest and most diverse protist groups. Like flagellates, many ciliates sit on the boundary between plantlike and animal-like lifestyles. The distinguishing anatomical characteristic of the group is a covering of fine hairlike cilia, which are embedded in the cell wall. A network of filaments and tubules (tiny threads and tubes) inside the cilia help coordinate movement and other activities. Most ciliates use

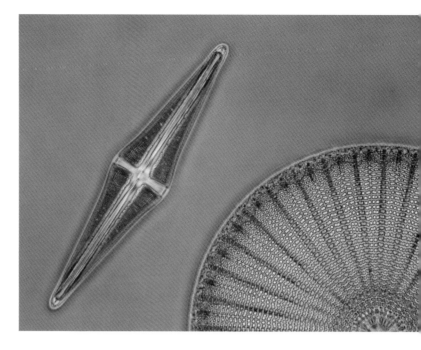

▲ *The diatom on the left has a typical pennate form. That on the right is a circular centric diatom.*

coordinated rhythmic beating of their cilia to "row" themselves through liquid.

● **Radiolarians** Radiolarians belong to the same large group of protists as the animal-like heliozoans and amoebas (the Sarcodina). Radiolarians are usually spherical and adorned with an assortment of fine needlelike spines radiating from the central cell body. They have a skeleton made of silicates or sulfates. Temporary outgrowths of the cell, called pseudopodia, stretch beyond the spines and gather food items. Many radiolarians contain photosynthesizing symbiotic algae.

● **Foraminiferans** Foraminiferans are among the most spectacular of plantlike protists. Foraminiferans are largely marine organisms and make use of the mineral calcium carbonate, which is abundant in seawater, to secrete an elaborate shell that usually grows in a spiral. Some are large enough to see with the naked eye.

● **Diatoms** Diatoms are easily identified by their uniquely structured cell walls. Unlike most plantlike organisms, which have cell walls made of cellulose, diatoms live within microscopic brittle boxes made from silica—the same mineral that forms sand and glass. The group is split into two distinctive types: centric and pennate diatoms. Centric diatoms have boxes that are circular, triangular, or polygonal (multisided) in flat view. Pennate diatoms have boxes shaped like pencil cases.

ANATOMY HIGHLIGHTS

EXTERNAL ANATOMY Each diatom cell is enclosed in an intricately sculpted two-part boxlike shell. The shell is made of silica, perforated with numerous pores, and often adorned with a variety of studs and spines. *See pages 166–167.*

INTERNAL ANATOMY A typical eukaryotic cell, with a large cell vacuole surrounded by cytoplasm containing the nucleus, endoplasmic reticulum, mitochondria, and one or more chloroplasts. *See pages 168–170.*

REPRODUCTIVE SYSTEM Routine reproduction is asexual, with each new generation slightly smaller than the last. Sexual reproduction also occurs, involving the production of sex cells by meiosis. Sexual reproduction results in a new, large individual. *See page 171.*

External anatomy

CONNECTIONS

COMPARE the plantlike diatom with the **AMOEBA**. Amoebas are animal-like in that they obtain nutrients by consuming organic matter, including other protists.

The epivalve *is the protective "lid" of the diatom.*

The frustule, *or theca, is the diatom's shell and cell wall.*

The raphe *is a furrow that runs along the middle of the epivalve in pennate diatoms.*

Many tiny areolae, *or pores, perforate the epivalve. They enable the diatom to absorb and excrete gases and nutrients.*

▶ TOP VIEW
Navicula
This is a view of the diatom from above, showing the epivalve. In life, the valve is around 0.001 inch (0.02 mm) long.

The most striking feature of diatoms is the cell wall. This is the means by which diatoms are usually classified. Unlike the walls of most other plantlike cells, which owe their strength mostly to the complex carbohydrate cellulose, the walls of diatoms contain up to 70 percent silica (silicon dioxide). The silica forms an intricately structured glassy shell called the frustule, or theca. The shell comes in two halves, called valves, which overlap one another like the top and bottom parts of a Petri dish or a pencil case. The valve that forms the "lid," the epivalve, is usually older and slightly larger than the bottom part, which is called the hypovalve. In pennate diatoms there is a conspicuous fissure or furrow called the raphe running along the length of the valve. The theca is perforated by thousands of pores, or areolae. The cell absorbs and excretes dissolved gases and nutrients through the areolae, which are just 0.1 to 0.6 micron across.

While the glassy silica frustule is a structure that is unique to diatoms, other plantlike protists also have distinctive outer coverings.

IN FOCUS

Useful structures

The fine structure and pores of diatom cell walls make them surprisingly useful to humans. The ultrafine pattern of ridges and studs on the shell of the marine pinnate genus *Pleurosigma*, for example, is used for testing the resolution of quality light microscopes—any imperfection in the lens will make the detail difficult to see. The remains of long-dead diatoms that accumulate in sediments at the bottom of oceans and lakes are also very useful. Known as diatomaceous earths, these sediments have a variety of industrial uses. Because of their high silica content they make excellent abrasives (like fine sand), and their tiny pores allow them to be used as ultrafine filters.

Dinoflagellates, for example, bear armorlike cellulose plates of varying thickness, and the exquisite radiolarians have a fine latticelike skeleton made of silicate or sulfate minerals, with delicate radiating spines. Radiolarians and the related heliozoans also bear fine needlelike projections known as axopods. These are extensions of the cell membrane filled with thick cytoplasm to give them rigidity. Foraminiferans secrete calcium carbonate, silicates, cellulose, or proteins, creating a series of chambers of increasing size as the organism grows. Other plantlike protists have flexible cell coverings, but these vary greatly in structure and function. The flexible cell wall of a ciliate such as *Paramecium* is highly complex. The cell wall has specialist organelles used for locomotion, feeding, and defense embedded in its structure.

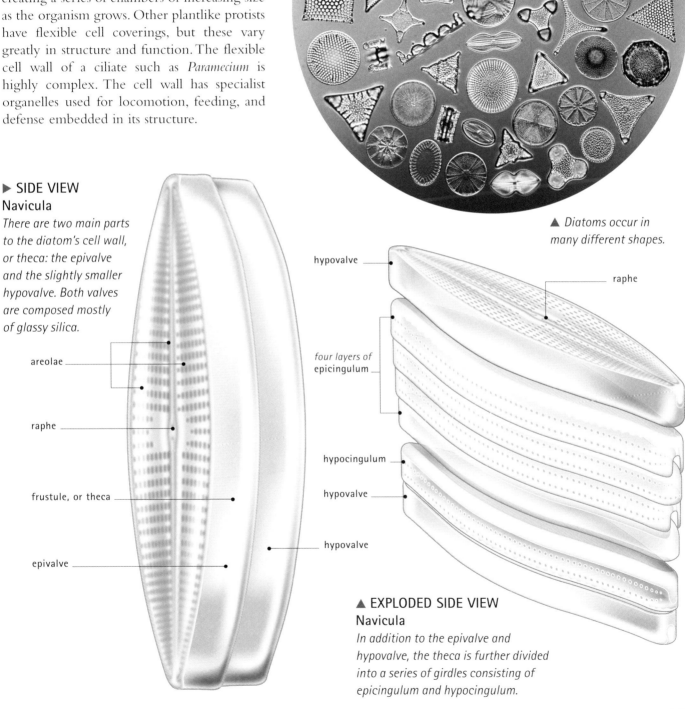

▲ Diatoms occur in many different shapes.

▶ **SIDE VIEW**
Navicula
There are two main parts to the diatom's cell wall, or theca: the epivalve and the slightly smaller hypovalve. Both valves are composed mostly of glassy silica.

areolae

raphe

frustule, or theca

epivalve

hypovalve

raphe

four layers of epicingulum

hypocingulum

hypovalve

hypovalve

▲ **EXPLODED SIDE VIEW**
Navicula
In addition to the epivalve and hypovalve, the theca is further divided into a series of girdles consisting of epicingulum and hypocingulum.

Internal anatomy

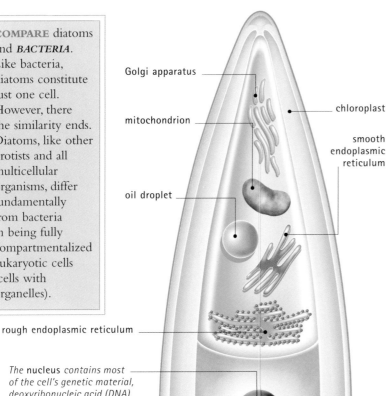

Golgi apparatus

mitochondrion

oil droplet

chloroplast

smooth endoplasmic reticulum

rough endoplasmic reticulum

*The **nucleus** contains most of the cell's genetic material, deoxyribonucleic acid (DNA). DNA molecules carry the genetic information necessary for organizing the cell.*

vacuole

frustule

*The **rough endoplasmic reticulum** is involved in making proteins.*

*The **smooth endoplasmic reticulum** is involved in making proteins.*

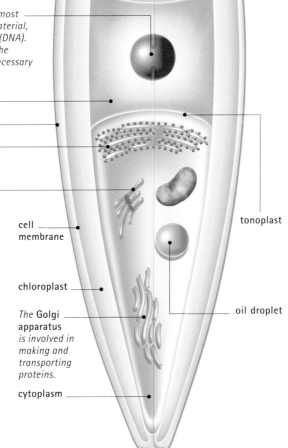

cell membrane

chloroplast

*The **Golgi apparatus** is involved in making and transporting proteins.*

cytoplasm

tonoplast

oil droplet

▶ ORGANELLES
Cross section of Navicula
All the living matter of a diatom is contained within the frustule. The chloroplasts convert sunlight, water, and carbon dioxide into sugars to fuel the cell. The nucleus and chloroplasts are all called organelles.

In single-celled organisms like diatoms, there are no internal organs as such, but there are various microscopic components that perform different functions in much the same way. These are known collectively as organelles, meaning "miniorgans."

On the inside, diatoms are fairly standard plantlike cells, with a large cell vacuole and a narrow peripheral area of cytoplasm. The peripheral area contains all the major organelles. The vacuole is separated from the cytoplasm by a membrane called the tonoplast. The outer cell membrane lies next to the intricate silica walls, tracing every curve and indentation of the ornate structure. The chemical building blocks for the wall are made inside the cell, within large secretory vesicles (small, enclosed compartments). The vesicles move to the cell membrane and merge with it to empty their contents outside. These building blocks rearrange themselves to form the frustule, or cell wall, while the cell membrane reseals below its new covering.

The position of the cell nucleus varies, but in most diatoms it is more or less central. The nucleus is often suspended in the central vacuole by strands of cytoplasm. The nucleus is

IN FOCUS

Making a move

Unlike many other protists, diatoms do not have cilia or flagella that help them swim. Diatoms living in the ocean or other bodies of open water float at the mercy of tides and currents. Pennate diatoms, however, are able to move as long as they are in contact with a surface. They glide on a stream of mucilage (slime) that they secrete from one end of the raphe (the groove in the cell wall). The mucilage flows along the raphe, allowing the diatom to move as if it were on tracks. Centric diatoms do not have this slime groove and so cannot glide.

Diatom origins

Diatoms make excellent fossils because their inorganic silica shells are hard and do not decay. Some ancient marine or lake deposits consist almost entirely of diatom remains. The oldest of these date to the late Paleozoic era, suggesting that diatoms first appeared more than 250 million years ago. The early examples were all centric diatoms that had flagellated male sex cells.

associated with a system of flat membrane-bound vesicles, such as the Golgi apparatus and the rough endoplasmic reticulum. In these vesicles, instructions coded in the genetic material of the nucleus are used to make the chemicals the cell needs to function, such as enzymes and structural proteins.

Like all eukaryotic cells, diatoms contain organelles known as mitochondria, in which the chemical energy available in sugars and fat is converted into a more convenient form, much as power plants convert fossil fuels or nuclear energy into electricity. Like power plants, mitochondria are dependent on a ready source of fuel. In most diatoms, as in plants, this

is provided by the largest and most conspicuous organelles in the cell, the chloroplasts. These are the microscopic factories where sunlight, water, and carbon dioxide are converted into sugars to fuel the cell. Above all else, it is the possession of chloroplasts and the ability to synthesize their own food that allies diatoms more closely with plants than with animals. Like green plants, diatom chloroplasts contain the green pigment chlorophyll. They also contain the yellow, gold, and brown pigments beta-carotene, diatoxanthin, diadinoxanthin, and fucoxanthin.

The products of photosynthesis may be transported around the cell for immediate use, or they may be converted to starch or fat for storage. Diatoms store the sugary products of photosynthesis in an unusual form. Most plants turn their sugars into starch (the compound that gives potatoes their floury texture). Diatoms, however, build the sugars into a polysaccharide called chrysolaminarin. Chrysolaminarin forms oily droplets inside the cells that are visible under a microscope.

Internal skeleton

As well as their hard mineral wall, diatom cells also have an internal "skeleton," similar to that found in other protists and in the cells of larger organisms. This cytoskeleton is made of very fine protein filaments called microtubules,

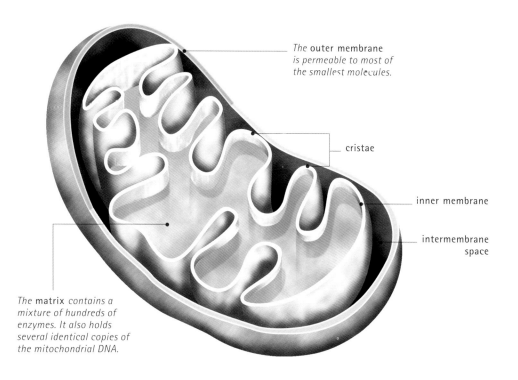

▶ **MITOCHONDRION**
This is an artist's conception of a mitochondrion. Each diatom contains membrane-bound organelles called mitochondria. These convert energy from chemical fuels such as sugars, starches, and fats into a form able to drive reactions within the cell.

*The **outer membrane** is permeable to most of the smallest molecules.*

cristae

inner membrane

intermembrane space

*The **matrix** contains a mixture of hundreds of enzymes. It also holds several identical copies of the mitochondrial DNA.*

which can be assembled and disassembled very rapidly to provide support and a transport network. Microtubules play an important part in cell division. They form a structure called the spindle. This apparatus helps guide the movement of chromosomes (structures that contain DNA) as they separate and multiply to form two new sets of genetic material. The cytoskeleton is also involved in locomotion: many plantlike protists, including the male sex cells of some diatoms, use flagella to move around. These flagella are oarlike or propeller-like structures and are supported by an internal scaffold of microtubules.

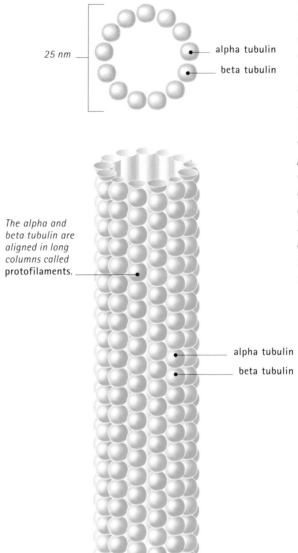

25 nm — alpha tubulin
beta tubulin

The alpha and beta tubulin are aligned in long columns called protofilaments.

alpha tubulin
beta tubulin

◄ MICROTUBULE
The molecules within a microtubule are arranged in a ring. The ring consists of alternating molecules called alpha and beta tubulin. These molecules are aligned in 13 long parallel rows, forming a cylinder. Microtubules are important in cell division; they provide internal support and a network for the transport of molecules through the cell.

EVOLUTION

The great divide

Probably the most fundamental division of life on Earth is that between prokaryotes and eukaryotes. The prokaryotes (bacteria) lack membrane-bound organelles such as mitochondria and chloroplasts, and their genetic material is loose in the cytoplasm. Eukaryotes have membrane-bound organelles, including a nucleus. Current thinking is that simple eukaryotic life arose from symbiotic relationships between prokaryotes, where one cell came to live inside another and performed a particular function that benefited the host. This theory explains the origins of cell organelles and why mitochondrial organelles have their own DNA.

▶ *Organelles are visible within some of these colonial centric diatoms. Although their valves are not touching, they are aligned top to bottom. Several species of diatoms, both centric and pennate, live together in colonies.*

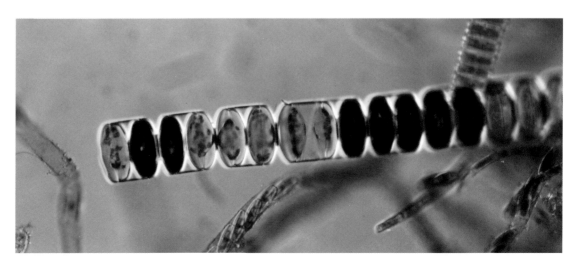

Reproductive system

Diatoms reproduce asexually most of the time. Each cell replicates vital organelles, such as the nucleus, mitochondria, and chloroplasts; separates the two parts of the cell wall; and splits in half so that each new "daughter cell" receives one valve and a working set of other cell components. Each of the new cells then grows a new base valve before they separate. Over time, this inevitably results in a shrinking lineage of cells—the base valve is always smaller than the "lid," and so the daughter that receives the bottom half will be slightly smaller than the one that gets the top.

Diatoms sometimes reproduce sexually, and this restores the lineage to full size. The nucleus of a sexually reproducing cell undergoes a form of double division called meiosis, which results in the production of male and female sex cells, each containing half the usual number of chromosomes. One male sex cell fuses with a female sex cell from another individual to make a fertilized cell called a zygote, which has a full set of chromosomes. The new individual grows rapidly, becoming an auxospore. A mature auxospore is a first-generation diatom, large enough to begin the cycle of asexual reproduction all over again.

The male sex cells of centric diatoms bear a long whiplike flagellum called a flimmer, which they use to swim. In most other protists, the flagella are supported by an internal scaffold of microtubules (the axoneme) in a ring of nine around a further two. This nine-plus-two arrangement is one of the most commonly repeated structures in nature. Diatoms, however, are different. Their flimmer flagella have a nine-plus-zero organization, lacking the central pair of microtubules.

AMY-JANE BEER

FURTHER READING AND RESEARCH
Harold, Franklin. 2001. *The Way of the Cell: Molecules, Organisms, and the Order of Life.* Oxford University Press: Oxford, UK.
Madigan, Michael T., J. M. Mantinko, and Jack Parker. 2002. *Brock's Biology of Microorganisms.* Prentice Hall: Upper Saddle River, NJ.

▶ **ASEXUAL REPRODUCTION**
When a diatom reproduces asexually it divides in two. One "daughter cell" takes the larger epivalve, and the other cell takes the hypovalve. Those cells that inherit the hypovalve are smaller than the parent cell.

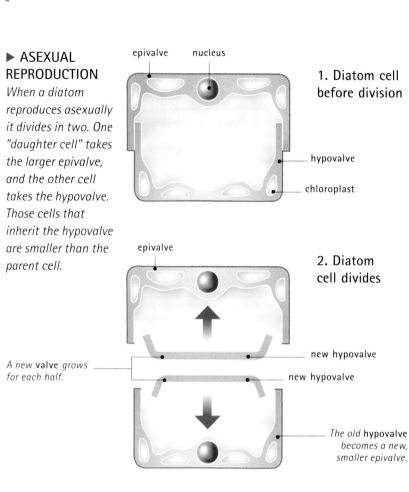

1. Diatom cell before division

2. Diatom cell divides

A new **valve** grows for each half.

new hypovalve

new hypovalve

The old **hypovalve** becomes a new, smaller epivalve.

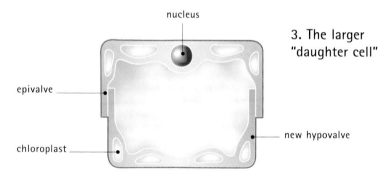

3. The larger "daughter cell"

epivalve

chloroplast

new hypovalve

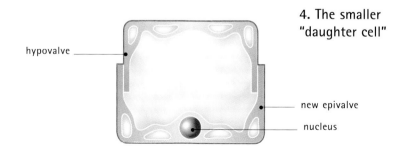

4. The smaller "daughter cell"

hypovalve

new epivalve

nucleus

Digestive and excretory systems

Digestion includes all the ways that Earth's diverse life-forms break down the food they eat into useful and vital products such as proteins, vitamins, minerals, sugars, and starches. The products of digestion fuel life processes, store energy, maintain health, fight disease, and provide the basic building blocks of life.

Plants, seaweeds, single-celled algae such as those in plankton, and some bacteria make their own food from chemicals around them. Plants use the energy of sunlight, gases in the air, and water to make sugars, so they do not need to digest food. Neither do many types of bacteria. Nitrogen-fixing bacteria combine nitrogen with hydrogen, using the energy released by the reaction to produce sugars. Nearly every other type of life-form, however, must eat and digest food to live.

Digesting food creates waste materials. Some products of digestion are harmful toxins; others are of no use to the organism. These wastes are dealt with and removed from the body by excretory systems.

Animal digestion

The complexity of animal digestive systems varies greatly. Specialized systems may be devoted to digesting certain foods, such as the many-chambered tracts of plant eaters. By contrast, gutless tapeworms simply absorb another animal's food. In vertebrates, the digestive system includes all the structures and organs along which food passes through the body.

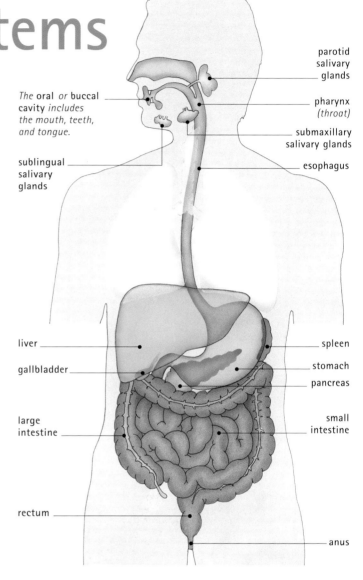

The oral or buccal cavity includes the mouth, teeth, and tongue.

parotid salivary glands
pharynx (throat)
submaxillary salivary glands
esophagus
sublingual salivary glands
liver
gallbladder
large intestine
rectum
spleen
stomach
pancreas
small intestine
anus

▲ DIGESTIVE SYSTEM
Human
The digestive system of all vertebrates, including humans, comprises the digestive tract and digestive glands. The digestive tract is a tubular passageway that extends from mouth to anus (or cloaca). Digestive glands line the walls of the tract. They include the pancreas, salivary glands, and the liver. All secrete enzymes that help digest food.

SYSTEM HIGHLIGHTS

MOUTH, TEETH, AND TONGUE The mouth, teeth, and tongue begin the process of breaking down food. *See pages 173–175.*

THROAT TO STOMACH The stomach is a stretchy bag containing acids and digestive juices that aid digestion. *See pages 176–179.*

INTESTINES Molecules of digested food, such as sugars and proteins, are absorbed into the bloodstream across the walls of the intestines. *See pages 180–183.*

LIVER, PANCREAS, AND GALLBLADDER The liver cleanses the blood, converting toxins into less harmful products. *See page 184.*

EXCRETORY SYSTEM The kidneys or kidney-like structures filter waste products. *See pages 185–187.*

CONNECTIONS

COMPARE how a *MUSHROOM* absorbs its food from organic matter with how plants make food and with how animals digest food.

COMPARE the ways in which *BACTERIA* and a single-celled algae such as a *DIATOM* feed.

Mouth, teeth, and tongue

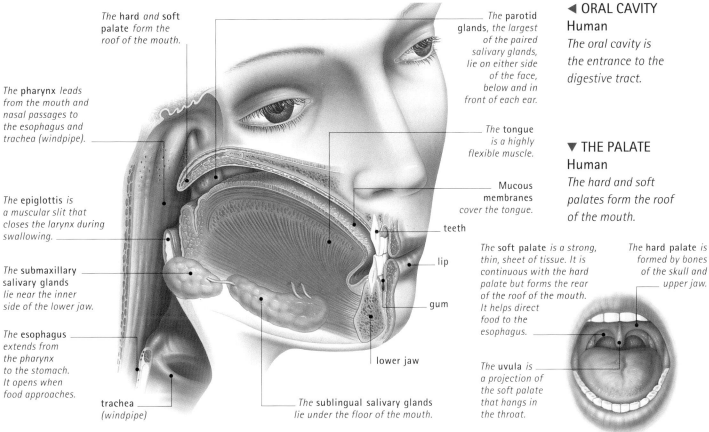

The **hard** *and* **soft palate** *form the roof of the mouth.*

The **pharynx** *leads from the mouth and nasal passages to the esophagus and trachea (windpipe).*

The **epiglottis** *is a muscular slit that closes the larynx during swallowing.*

The **submaxillary salivary glands** *lie near the inner side of the lower jaw.*

The **esophagus** *extends from the pharynx to the stomach. It opens when food approaches.*

trachea (windpipe)

The **parotid glands**, *the largest of the paired salivary glands, lie on either side of the face, below and in front of each ear.*

The **tongue** *is a highly flexible muscle.*

Mucous membranes *cover the tongue.*

teeth

lip

gum

lower jaw

The **sublingual salivary glands** *lie under the floor of the mouth.*

◀ ORAL CAVITY
Human
The oral cavity is the entrance to the digestive tract.

▼ THE PALATE
Human
The hard and soft palates form the roof of the mouth.

The **soft palate** *is a strong, thin, sheet of tissue. It is continuous with the hard palate but forms the rear of the roof of the mouth. It helps direct food to the esophagus.*

The **hard palate** *is formed by bones of the skull and upper jaw.*

The **uvula** *is a projection of the soft palate that hangs in the throat.*

The vertebrate digestive tract starts with the mouth, teeth, and tongue. These make up the oral, or buccal, cavity. The oral cavity's main task is to prepare food for its journey through the body. In mammals, fleshy lips also allow young to suckle, since they enable an airtight seal, and the mouth, teeth, and tongue are often vital for communicating.

The teeth, lips, inner cheeks, hard palate, and tongue of a mammal all help masticate (chew) food and form it into a rounded mass, or bolus, for swallowing. Only mammals chew. Other animals use their oral structures to catch, hold, or tear up prey. Mastication breaks the food into smaller pieces and mixes it with saliva, which moistens and lubricates the bolus for its journey down the throat. These tasks are mechanical. Saliva also starts the process of chemical digestion. It begins to break down food. The saliva of some mammals, including humans, contains a starch-digesting enzyme, amylase. However, many meat eaters, such as cats and dogs, have no amylase in their saliva, since their diet contains very little starch.

Teeth
A few invertebrates have structures that function as teeth, but only vertebrates have true teeth made from minerals like enamel (in mammals) and dentine. The crown is the visible part above the gum; the root is the unseen part, attached to the jaw by fibers.

Teeth vary a great deal, and their differences are among the criteria used by biologists to help them classify different groups of animals. Many vertebrates have a mouthful

▶ MOLAR TOOTH
Mammal
The crown of a mammal's tooth has an outer layer of enamel, the hardest tissue in the body; softer cementum covers the root. Bonelike dentine forms much of the tooth's interior. Blood vessels and nervous tissue make up the innermost pulp.

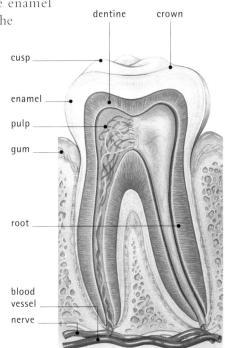

dentine   crown

cusp

enamel

pulp

gum

root

blood vessel

nerve

of similar-looking teeth. Sharks and other predators, for example, tend to have pointed teeth that can pierce and hold prey. Many fish have fused teeth perfect for scraping food from surfaces. Snakes have backward-curving teeth that grip prey. Turtles and birds have no teeth. Mammals have specialized teeth such as incisors, canines, and molars. The tusks of elephants and walruses are teeth, too. Nearly all vertebrates are born with one set of teeth that are replaced as they wear out.

Most mammals have only two sets of teeth: primary (or first, baby, or milk) teeth, which are replaced by permanent, or adult, teeth. Primary teeth tend to be smaller and more prone to wear than adult teeth. With a few exceptions, adult teeth are not replaced and no longer grow once in place. The incisors of lagomorphs (hares and rabbits) and rodents are worn down by gnawing; these teeth continue to grow from the root throughout the animal's life. An elephant's molars emerge in stages, pushing older molars to the front of the tooth row. Elephants' molars are replaced from behind until quite late in the elephant's life when all the molars have emerged.

Incisors, canines, and molars

Mammals have four main types of teeth: incisors at the front of the mouth, canines on either side of the incisors, premolars along the side of the mouth, and molars at the back. The number and shape of each type vary according to species. Incisors are chisel-shaped and used for cutting and clipping food. Rodents and lagomorphs have long incisors that can gnaw tough plant materials. Canines are pointed teeth that meat eaters use for puncturing, holding, and tearing food. Premolars and molars are large teeth with ridged or peaked surfaces called cusps. They crush and grind food. The fourth upper premolars and first lower molars of carnivores, such as bears, cats, and dogs, are called carnassials. They are used for slicing into the flesh of prey.

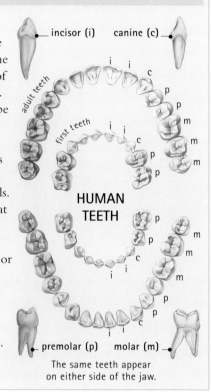

HUMAN TEETH

The same teeth appear on either side of the jaw.

Carnivore, omnivore, herbivore, or bloodsucker?

An animal's teeth meet the requirements of its diet. Carnivores (meat eaters) need sharp teeth that can pierce skin and hold onto prey. The teeth of some carnivores, including snakes and many fish, curve or point backward. Such teeth act as a cage to hold prey in the mouth and direct it toward the throat. Herbivores (plant eaters) need broad, crushing teeth to chew tough leaves and stems. They also need teeth such as incisors to clip leaves from trees or to pull up grass stalks. Omnivores need various types of teeth to deal with their mixed diet.

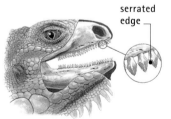

An iguana has peglike teeth with crowns that form a leaf shape. The edges are serrated (jagged) for cutting plants and piercing small prey.

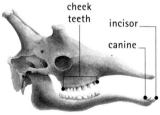

A giraffe has no upper incisors but uses its long tongue and upper lip to pull leaves from trees. Cheek teeth have crescent-shape cusps for grinding.

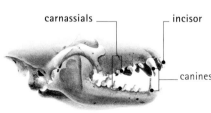

A wolf has large, pointed canines, upper incisors, and carnassial cheek teeth—all perfect for gripping, tearing, and slicing living animal prey.

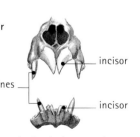

A vampire bat's sharp canines and incisors slice skin. Cheek teeth have few crushing surfaces, since the bat has a liquid diet of vertebrate blood.

Spider mouthparts

Spiders have two pairs of feeding appendages. The chelicerae bear fangs that deliver venom. The sensitive pedipalps are used to manipulate food.

Those of males are also used for sperm transfer. The pedipalps often bear enlarged lobes called maxillae, which are used to crush and grind food.

1. *A spider buries its fangs in its prey, injecting paralyzing venom. The chelicerae move either forward and down, or together sideways.*

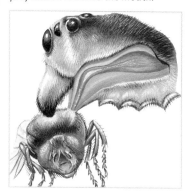

2. *The spider regurgitates fluids onto the prey. The fluids contain enzymes that begin to digest the prey before it enters the mouth.*

3. *The spider sucks its food into its mouth. Some spiders rip larger prey into shreds with epidermal teeth on their chelicerae before feeding.*

Not all animals, or even all vertebrates, have true teeth. Epidermal teeth are hard, horny projections that function as teeth. Tadpoles, adult platypuses, and jawless fish such as lampreys all have epidermal teeth.

Invertebrate mouthparts

The oral structures of invertebrates vary greatly. Mollusks, such as octopuses, snails, and giant clams, have a hard, tonguelike radula that they use to grind down prey. Corals, hydras, and sea anemones have an opening to their digestive cavity that serves as both mouth and anus. Flatworms, are similar, and also have a series of excretory pores dotted around their body through which wastes are voided. Other invertebrates have a complete digestive tract—that is, a tube with two openings: the mouth and the anus.

Arthropods are invertebrates with jointed limbs and a hard outer exoskeleton. The group includes insects, crustaceans (such as crabs, shrimp, and lobsters), and spiders. Many arthropods have paired, jointed, and limb-like appendages or mouthparts called, for example, mandibles, maxillae, chelicerae, and maxillipeds. These appendages are mechanical digesters. They catch, filter, hold, snip, tear up, and generally manipulate food before passing it to the mouth. The form, arrangement, and numbers of each appendage vary immensely between different groups. In lobsters and crayfish the first legs act as mouthparts, too. Either the left or the right claw is massive and has epidermal teeth for crushing hard-shelled prey such as snails and clams.

Filter feeders

Not all animals need to break up large food items. Many animals filter tiny organisms from their watery homes and swallow them whole. Arthropod filter feeders often use appendages (such as maxillae) to filter food from the water. Small aquatic crustaceans called copepods have tiny hairlike setae on their maxillae. Like paddles, these direct food particles to the mouth without touching them. Rorqual whales filter small animals from the water with thin plates of baleen that hang from their upper jaw.

▼ BEAK AND RADULA

Common octopus
An octopus's mouth has a strong sharp beak for breaking into hard-shelled prey, and a tonguelike radula covered in hard spikes that grind down prey.

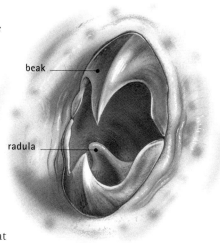

beak

radula

175

Throat to stomach

A vertebrate's foregut consists of the pharynx (throat), esophagus (tube joining pharynx to stomach), and the stomach. Invertebrate foreguts do not always include the stomach, which may be part of the midgut.

The pharynx and swallowing

The pharynx performs roles in the respiratory and digestive systems. In most bony fish it includes the gills, which are located in a series of slits leading from the pharynx to the exterior. In many terrestrial vertebrates the pharynx directs air into the trachea (windpipe) and food into the esophagus. The key digestion-related role of the pharynx is swallowing. Once begun, swallowing is a reflex that cannot be stopped. Most vertebrates bolt their food, swallowing it whole, and the esophagus expands to suit. Seabirds toss food to the back of their throat. In most mammals the back and sides of the tongue expand against the soft palate, forcing food into the esophagus. The epiglottis (a muscular slit) briefly closes the upper part of the trachea, the larynx, which contains the voice box. This prevents food from entering the trachea. In most mammals, the epiglottis forms a seal with the soft palate that keeps the food and air passages separate but open. In adult humans, however, the epiglottis and larynx are much lower, enabling speech. Therefore, the passages are kept separate, but only the esophagus is open during swallowing. That is why adult humans cannot breathe when they swallow. Infant humans can breathe and suckle, however, since their larynx has not yet descended, and the epiglottis–soft palate seal works as it does in other mammals.

The esophagus

The esophagus is a slender, muscular tube that expands around food. Cells in the lining secrete thick, sticky mucus, which greases the passage of food to the stomach. The esophageal lining of animals that eat rough or scratchy foods contains keratin, the key protein in hair, fur, and fingernails. Keratin hardens and protects the lining against damage. Very few esophagi secrete chemicals (enzymes) that digest food.

COMPARE the pharynx of a sea squirt or the foregut of a vertebrate with the crop of a *DRAGONFLY*.

COMPARE how a human swallows with how another mammal, such as the *ZEBRA*, swallows.

◄ PHARYNGEAL SPINES
Leatherback turtle
The throat of a sea turtle is lined with backward-pointing spines, which enable it to feed on jellyfish. The spines puncture the jellyfish and prevent the watery animal from slipping away.

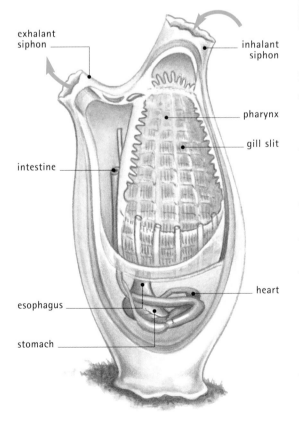

◄ PHARYNX
Sea squirt
A sea squirt's pharynx is a netlike basket full of tiny holes called gill slits. Particles of food are trapped by sticky mucus secreted by the pharynx. The mucus is rolled into a cord and passed to the intestine. The pharynx also bears gills, which remove oxygen from the water, thus enabling the animal to breathe. Water enters through the inhalant, or buccal, siphon and exits via the exhalant siphon.

exhalant siphon

inhalant siphon

pharynx

gill slit

intestine

heart

esophagus

stomach

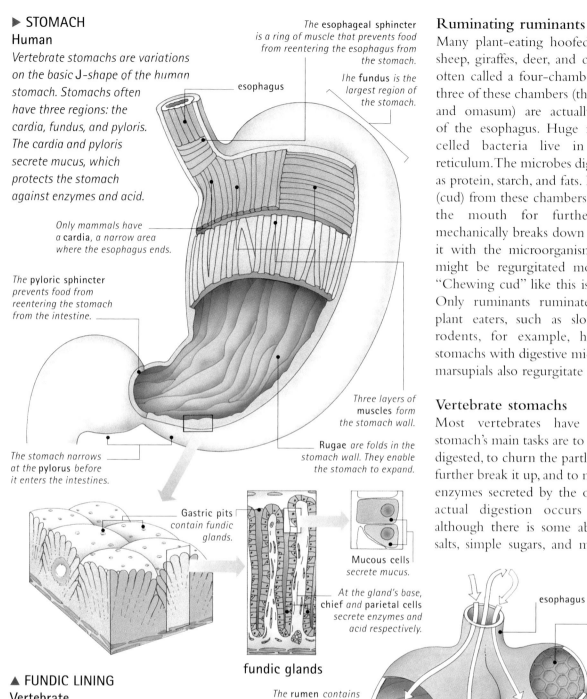

▶ STOMACH
Human

Vertebrate stomachs are variations on the basic J-shape of the human stomach. Stomachs often have three regions: the cardia, fundus, and pyloris. The cardia and pyloris secrete mucus, which protects the stomach against enzymes and acid.

*Only mammals have a **cardia**, a narrow area where the esophagus ends.*

*The **pyloric sphincter** prevents food from reentering the stomach from the intestine.*

*The stomach narrows at the **pylorus** before it enters the intestines.*

esophagus

*The **esophageal sphincter** is a ring of muscle that prevents food from reentering the esophagus from the stomach.*

*The **fundus** is the largest region of the stomach.*

*Three layers of **muscles** form the stomach wall.*

__Rugae__ are folds in the stomach wall. They enable the stomach to expand.

Gastric pits *contain fundic glands.*

__Mucous cells__ secrete mucus.

At the gland's base, **chief** *and* **parietal** *cells secrete enzymes and acid respectively.*

fundic glands

▲ FUNDIC LINING
Vertebrate

The lining of the fundus is dotted with fundic glands, which secrete mucus, hydrochloric acid, and digestive enzymes. The acid raises the pH (relative acidity) to levels at which the enzymes work best. The enzymes then begin to digest the food.

▶ STOMACH
Ruminant

The rumen, reticulum, and omasum are lined with the same tissues that line the esophagus, though differently arranged. These linings secrete mucus but not digestive enzymes. The abomasum is the "true" stomach. It secretes digestive enzymes. Ruminants are named for the largest of these four chambers, the rumen.

Ruminating ruminants

Many plant-eating hoofed mammals such as sheep, giraffes, deer, and camels have what is often called a four-chamber stomach. In fact, three of these chambers (the rumen, reticulum, and omasum) are actually enlarged regions of the esophagus. Huge numbers of single-celled bacteria live in the rumen and reticulum. The microbes digest cellulose as well as protein, starch, and fats. Partly digested food (cud) from these chambers is regurgitated into the mouth for further chewing. This mechanically breaks down the food and mixes it with the microorganism soup. One bolus might be regurgitated more than 40 times. "Chewing cud" like this is called ruminating. Only ruminants ruminate, but some other plant eaters, such as sloths, peccaries, and rodents, for example, have multichamber stomachs with digestive microorganisms. Some marsupials also regurgitate to re-chew.

Vertebrate stomachs

Most vertebrates have a stomach. The stomach's main tasks are to store food until it is digested, to churn the partly digested food and further break it up, and to mix it with digestive enzymes secreted by the organ's lining. Little actual digestion occurs in the stomach, although there is some absorption of water, salts, simple sugars, and minerals through its

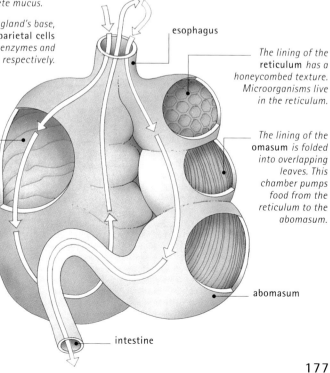

*The **rumen** contains microorganisms.*

esophagus

*The lining of the **reticulum** has a honeycombed texture. Microorganisms live in the reticulum.*

*The lining of the **omasum** is folded into overlapping leaves. This chamber pumps food from the reticulum to the abomasum.*

abomasum

intestine

177

Peristalsis

From esophagus to anus or cloaca, food is pushed through the digestive system by waves of muscular contractions collectively called peristalsis. This occurs in many animals, including invertebrates with tubular digestive systems. The digestive cavities of some invertebrates that do not perform peristalsis are lined with tiny filaments called cilia.

They beat to drive food through the gut. On occasion, peristalsis works in reverse. In birds that feed on fruits with waxy coatings, reverse peristalsis forces intestinal contents back into the gizzard for further grinding and, also, mixing with secretions from the intestine that help break down the fruits' tough skin.

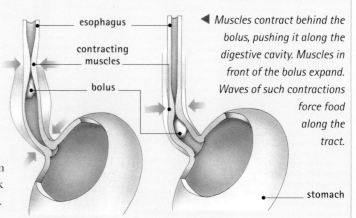

◀ *Muscles contract behind the bolus, pushing it along the digestive cavity. Muscles in front of the bolus expand. Waves of such contractions force food along the tract.*

walls. Some fish do not have a distinct stomach. A lamprey's diet, for example, consists of blood and flakes of tissue ripped from the body of a host. A lamprey does not need a storage organ; food passes straight from the esophagus to the intestines.

Invertebrate foreguts and midguts

The forms of invertebrate foreguts are as diverse as invertebrates themselves. Some, such as the sandworm or ragworm, do not have a stomach. Instead, esophageal pouches called ceca perform similar tasks. During feeding, the front of the foregut (the pharynx) is turned inside out and everted from the mouth. The inner walls of the pharynx bear teeth and jaws and are lined with tough, shiny chitin, which also lines the outer body. The pharynx retracts when prey is caught, dragging it inside.

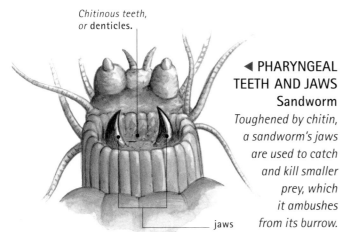

Chitinous teeth, or denticles.

◀ PHARYNGEAL TEETH AND JAWS
Sandworm
Toughened by chitin, a sandworm's jaws are used to catch and kill smaller prey, which it ambushes from its burrow.

jaws

Crops and gizzards

Many birds have an enlarged, thin-walled region of the esophagus called the crop. The crop stores food en route to the stomach and, in some birds, is used to carry food to the young. Also, a gizzard often occurs between the saclike crop and the intestines. The hind part of the stomach, the gizzard has thick muscular walls and contains small stones, or gastroliths, that have been deliberately swallowed by the bird. Gastroliths serve the same function as teeth in other vertebrates, since they crush and grind food, breaking it down mechanically. A few other animals, such as earthworms, alligators, and crocodiles, also have gizzards. Earthworms also have a crop. Some whales and dolphins have an enlarged region of their esophagus that acts like a bird's crop.

▶ DIGESTIVE SYSTEM
Pigeon
A pigeon secretes "milk" from the lining of the crop. This is fed to chicks until around 20 days after hatching.

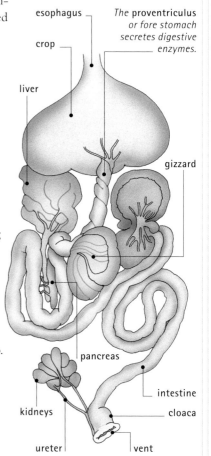

esophagus

crop

liver

The **proventriculus** *or fore stomach secretes digestive enzymes.*

gizzard

pancreas

kidneys

ureter

vent

intestine

cloaca

Arthropod foreguts and midguts

The foreguts of arthropods, including insects and spiders, also have a chitinous lining. Arthropods have an external skeleton called an exoskeleton. The exoskeleton's outer layer is the chitinous cuticle, which extends into and lines the foregut and hindgut (intestines). The midgut is not lined by cuticle.

Backward-pointing spines or "teeth" might project from the surface of an insect's foregut. An insect's muscular pharynx sucks and swallows, and the esophagus may be enlarged to form a crop. The crop empties into the "stomach," or mesenteron (midgut). Some insects have a muscular gizzard between crop and midgut. Invertebrate gizzards generally contain teeth or grinding plates rather than gastroliths, tiny stones that occur in the gizzards of birds and their ancestors, dinosaurs. Digestion and absorption of the products of the process occur in the insect midgut. The cells that line the midgut have folded surfaces that form fingerlike projections called villi. These are covered by even tinier bumps called microvilli. The projections increase the surface area across which the products of digestion can be absorbed. Midgut pouches called ceca may further increase the surface area.

A crustacean foregut might be a simple tube. However, the foreguts of crabs, lobsters, and other decapods (10-legged forms) contain a structure called the gastric mill. This is a series of hard plates that grind food. Hairlike setae prevent large particles of food from entering the midgut. The midgut often contains one or more diverticula, or pouches, inside which digestion occurs.

Mollusk foreguts and stomachs

Most filter-feeding bivalve mollusks, such as clams and oysters, have gills that pick up oxygen and obtain particles of food from the water. The complex stomach contains one end of a rodlike crystalline style, which rotates and grinds against a hard area of the stomach's wall. As it rotates it dissolves, releasing enzymes that begin to digest the filtered food. In snails and other nonbivalves the foregut includes glands and "teeth" made from tough chitin. These teeth, called the radula, bite, tear, and scrape food materials.

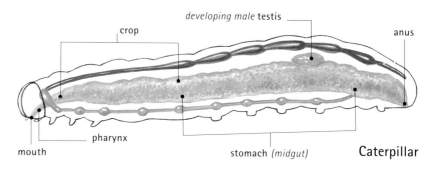

Caterpillar

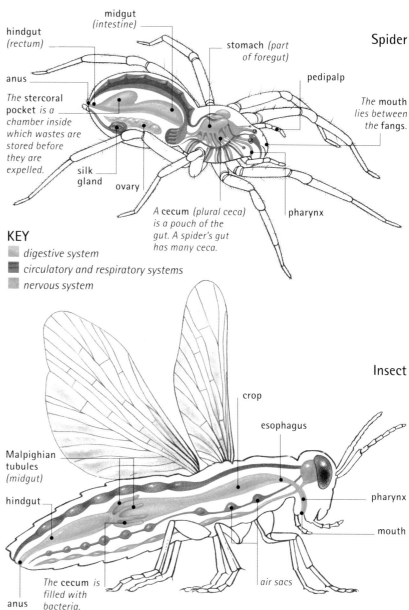

Spider

KEY

- digestive system
- circulatory and respiratory systems
- nervous system

Insect

▲ DIGESTIVE SYSTEMS
Arthropods

Caterpillars are eating machines, and their large crop, stomach, and intestines reflect this. The pharynx and stomach of a spider are attached to strong muscles. When the muscles contract, the pharynx and stomach dilate. This creates a vacuum that sucks in the spider's liquid food. The ceca of a plant-eating insect might be filled with microorganisms that help digest tough cellulose.

Intestines

In vertebrates, and many invertebrates, too, the intestines are where most chemical digestion takes place. This involves enzymes breaking down food into reusable products such as amino acids, the building blocks of proteins; starchy carbohydrates, which provide energy; and fatty acids. The intestines are also where most products of digestion are finally absorbed into the body. In vertebrates, the products are absorbed through the gut wall, passing into blood vessels in the cavity's lining. They are then carried to where they are needed in the body.

Structure of the digestive tract

From the esophagus to the anus (or cloaca), the digestive cavity of most vertebrates has a common structure. The walls have four layers. The innermost layer is called the mucosa. This includes the epithelium, which lines the cavity, the lamina propria, and the muscularis mucosa. Epithelia vary among species and among locations in the gut. Rodents that eat abrasive things such as insects, seeds, and grasses have epithelia strengthened in places with a tough protein called keratin.

Next to the epithelia lies the lamina propria, a sheet of smooth connective tissue. The muscularis mucosa is a layer of smooth muscle fibers. The muscularis mucosa is so named because it is muscular and because the epithelium contains cells that secrete mucus. Mucus is a thick, sticky fluid that, in the digestive system, aids the passage of food through the tract and protects the cells of the gut wall from the action of digestive enzymes and, in the stomach, acid.

Beneath the mucosa lies the submucosa. This consists of thick but loose connective tissue, and nerves that are part of the autonomic nervous system. This system controls the internal organs without any conscious effort by an animal. Of the sections of the gut, only the esophagus can be contracted at will.

The third of the layers of the mucosa is called the muscularis externa, which has an inner and outer layer of smooth muscle fibers. The inner

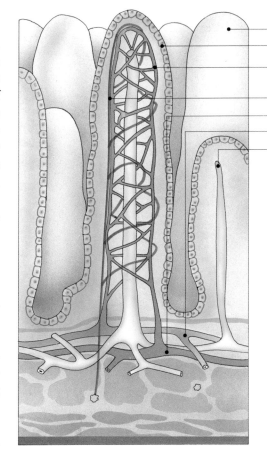

villus (plural, villi)
epithelial or mucous cell
capillary (tiny blood vessel)
nerve
artery
vein
A lacteal, a lymph vessel that absorbs essential fats.

◄ VILLI
Vertebrate
The inner lining, or mucosa, of the small intestine bears tiny fingerlike projections called villi. Even smaller projections called microvilli (not shown) form a rough "brush border." One epithelial cell might have as many as several thousand microvilli. Villi and microvilli increase the surface area available for absorption.

▼ DIGESTIVE TRACT
Vertebrate
The four layers of mucosa, submucosa, muscularis externa, and adventitia (or serosa) are common to the whole digestive tract, from esophagus to anus or cloaca. Glands occur in the mucosa and submucosa.

The epithelium is covered with villi.

serosa *or* adventitia
muscularis externa
outer longitudinal fibers
inner circular fibers
submucosa
nerve

mesentery
blood vessels
Folds called the plicae circulares *occur in the small intestine.*
mucosa
muscularis mucosa
lamina propria
epithelium

180

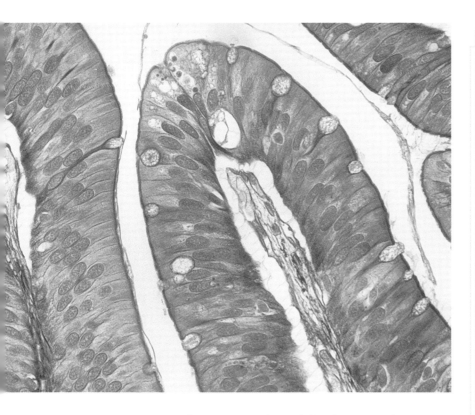

▲ *The gray dots on this cross section from a gut wall are goblet cells. They secrete mucus, a vital fluid that lubricates food as it passes along the gut and helps protect the gut walls from acids and enzymes.*

CONNECTIONS

COMPARE the intestines of a cartilaginous fish such as the *HAMMERHEAD SHARK* with the intestines of a bony fish such as a *SEA HORSE*.

COMPARE the intestines of a carnivore such as a *WOLF* with the intestines of a herbivore such as a *RED DEER* and an omnivore like a *GRIZZLY BEAR*.

layer is circular, while the outer layer runs lengthwise. The fourth, and outer, mucosa layer is the tunica adventitia, which is formed by loose, fibrous connective tissue. In places, the adventitia includes the mesentery. This thin membranous tissue envelops and supports the intestines. As well as an extensive web of blood vessels that carry blood to or from the organs it surrounds, the mesentery contains fatty areas that keep organs warm. Where a mesentery is present, the outer layer of the digestive tract is called the serosa.

Lymphoid tissue occurs throughout the layers of the gut wall, both as nodules and as larger lymph vessels. Lymph is a pale watery fluid containing white blood cells called macrophages and lymphocytes. These cells attack and destroy foreign particles in the body. Fish do not have lymph tissue in their gut wall, but all other vertebrates do.

The small intestines

The intestines of most vertebrates have two main regions: small and large. The terms refer to the diameter of the cavity rather than to its length. The small intestine can be long but is narrower than the large intestine. The duodenum, jejunum, and ileum are the three regions of the small intestine. Soupy chyme

Short or long?

The length of an animal's intestines is often related to diet. Herbivores must digest tough plant material to extract nutrients. A long intestine ensures that food takes a long time to travel through the gut, thus maximizing the time spent on digestion. In herbivores, blind pockets called ceca often occur at the junction of large and small intestines. Ceca further increase the length of the digestive system. Cellulose-digesting microorganisms often live inside ceca.

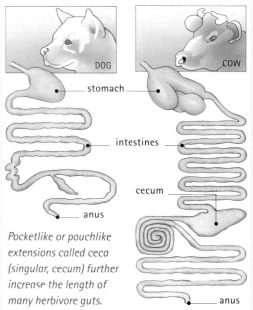

CARNIVORE   HERBIVORE

Pocketlike or pouchlike extensions called ceca (singular, cecum) further increase the length of many herbivore guts.

Terrestrial carnivores that eat only meat tend to have relatively short guts. Dolphins have relatively long intestines, for carnivores. Sometimes guts are so short, however, that they have evolved specializations to prolong the amount of time food spends in the digestive system. A spiral valve in the digestive cavity of some fish increases the amount of time food spends in their otherwise straight gut by forcing the bolus of food to take a winding path. Lampreys, sharks, lungfish, and sturgeon are among the fish that have spiral valves inside their intestines. Perch and other bony fish generally have no spiral valve but longer, coiled intestines.

Hindgut fermenters

In nonruminant plant-eating mammals such as rabbits and horses, the cecum is a wide branch at the start of the large intestine. Millions of plant-digesting microorganisms live in such a herbivore's cecum. By the time the bacteria have done their job of breaking down the tough plant material, the food has already passed the small intestine. The animal eats partly digested droppings from the cecum to ensure that the small intestine can absorb nutrients such as vitamin B_{12}. Droppings from the cecum are soft and coated in mucus, unlike the dry fully digested pellets. This process is called hindgut fermentation, since the bacteria ferment the food to make sugars as they digest it. Ruminants are foregut fermenters.

▼ *A jackrabbit eats its droppings, a form of feeding called coprophagy.*

(partly digested food) from the stomach enters the duodenum. Duodenal, or Brunner's, glands in the submucosa release chemicals that neutralize the acidic chyme. Other organs release substances into the duodenum. They include the liver, which releases bile via the gall bladder. Bile helps digest fats. The pancreas releases protein-splitting enzymes. The jejunum and ileum are differentiated by the types and structures of their mucosal glands but have no clear demarcation.

The epithelium of the small intestine bears fingerlike villi and microvilli. Villi and microvilli provide a calm, warm spot where chemical digestion can better take place away from the large central cavity. The shape of villi vary in different parts of the small intestine. In the human duodenum, for example, the villi are large, closely packed, and often leaf-shaped. Villi and microvilli also ensure a large surface for absorption, along with folds in the mucosa called plicae circulares. These folds are present throughout the small intestine except in the first portion, or bulb, of the duodenum and in the lower part of the ileum.

Large intestines

Further absorption occurs in the large intestine, where water especially is drawn back into the body. The large intestine has microvilli but neither plicae circulares nor villi. It joins with the small intestine at a sphincter (ring of muscle) called the ileocolic valve and leads to the anus or cloaca.

The first section of the large intestine is called the cecum. Joined to the cecum is a narrow, blind-ending tube called the appendix. It is a structure unique to humans and apes. In human's distant ancestors it played a role in digestion, but now has no function; it is vestigial. Wombats, civets, rabbits, and many other animals have an appendix-like structure, but these evolved from the cecum independently and are not analogous to the human version.

In mammals the cecum leads to a large hanging portion of the large intestine called the colon. This in turn leads to the final section of the large intestine: the rectum. The rectum narrows to form an anal canal. The expulsion of waste material from the anal canal is controlled, consciously, by a sphincter.

▶ *Tapeworms have no gut. They do not need a digestive cavity, since they are parasites that live inside the guts of vertebrate hosts. A tapeworm absorbs the products of its hosts' digestive system across its skin. This tapeworm lived inside a sheep.*

COMPARE the gut of an insect such as an *ANT* with the intestines of a crustacean such as a *CRAB* or *LOBSTER*, a mollusk such as a *GIANT CLAM*, or a cnidarian such as a *JELLYFISH*.

CONNECTIONS

Invertebrate intestines

Chemical digestion and absorption in invertebrates take place mostly in the intestines. An earthworm, for example, has a relatively simple tubular digestive system. Food passes through the esophagus, is then ground up and mixed in the gizzard, and is stored in the crop until digestion. After that the food is ready to enter the intestines, where digestion is completed and absorption into the body takes place. The intestine's epithelial cells secrete digestive enzymes. As in vertebrates, water is absorbed toward the end of the intestines, and waste exits via the anus.

In insects, microvilli occur in the stomach, which forms part of the midgut, so chemical digestion and absorption do occur in the stomach as well as the hindgut, or intestines. Also, many insects inject saliva into or onto their food before eating, so a considerable amount of chemical digestion takes place before food is even eaten. Similarly, spiders also begin chemically digesting their food before it enters the mouth.

The insect hindgut is generally divided into the pylorus, ileum, and rectum. The pylorus connects the midgut and the hindgut. It sometimes forms a valve. Generally, the ileum is a narrow tube that leads to the rectum. In some insects, the rear of the ileum differs sufficiently to be called a colon. The insect hindgut has a smooth chitinous lining but the layer of tissue (the apical plasma membrane) beneath this cuticle is extensively folded. Like vertebrate villi, these folds increase the surface area available for digestion and absorption.

Like herbivorous vertebrates, the intestines of invertebrates often contain colonies of microorganisms. While some insects have colonies in midgut ceca, others have them in their ileum. Larval (young) scarab beetles have an enlarged microorganism-containing ileum called a fermentation chamber, and termites have one called a paunch. The microorganisms enable the insects to digest tough plant material, even wood.

The insect rectum contains pads that reabsorb water. The rectal chamber may have other functions. Dragonfly nymphs develop in freshwater. Their rectum is lined by gills. They pump water over their rectal gills to enable respiration. They also use the pumping muscles to force water from the chamber, jet-propelling the insect through the water.

CLOSE-UP

Invertebrates without intestines

Many invertebrates do not have a digestive cavity with a separate foregut and hindgut. Hydras are small aquatic invertebrates related to sea anemones and corals. They have a simple canal-like cavity inside their body. The entrance serves as both mouth and anus. The products of digestion absorb directly into body tissues. Corals, sea anemones, and jellyfish have similar guts. Planarian flatworms have a more complex, branching gut. Food is broken down by enzymes, then absorbed by cells lining the digestive cavity and passed to the bloodstream. The branches of the flatworm's gut deliver the products of digestion to the whole body. Wastes exit through the mouth or through excretory pores. Tapeworms have no gut at all. They absorb the products of their host's digestion and so do not need a gut of their own.

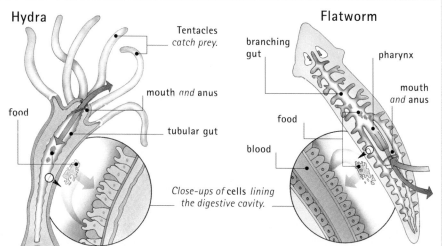

Hydra

Tentacles *catch prey.*

mouth *and* anus

food

tubular gut

Flatworm

branching gut

pharynx

mouth *and* anus

food

blood

Close-ups of **cells** *lining the digestive cavity.*

Cells that line the digestive cavity of a hydra digest food both externally (by releasing enzymes) and internally by engulfing particles for digestion within the cell.

A flatworm's mouth is on the underside of its body. The pharynx extends out of the mouth like a tube during feeding. Blood carries digested food products to body tissues.

183

Liver, pancreas, and gallbladder

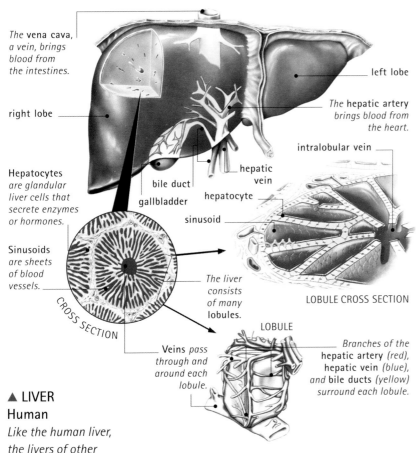

The **vena cava**, a vein, brings blood from the intestines.

right lobe

Hepatocytes are glandular liver cells that secrete enzymes or hormones.

Sinusoids are sheets of blood vessels.

CROSS SECTION

left lobe

The **hepatic artery** brings blood from the heart.

intralobular vein

hepatic vein

bile duct

hepatocyte

gallbladder

sinusoid

The liver consists of many lobules.

LOBULE CROSS SECTION

Veins *pass through and around each lobule.*

LOBULE

Branches of the **hepatic artery** *(red),* **hepatic vein** *(blue), and* **bile ducts** *(yellow) surround each lobule.*

▲ LIVER
Human

Like the human liver, the livers of other vertebrates are housed within the rib cage. The liver's shape depends on the shape of the animal. A snake's liver is long and narrow, for example. Most livers have two lobes, left and right. The microscopic structures of a liver do not vary, however. Sheets of hepatocytes are interspersed with sinusoids (sheets of blood vessels). Venous blood (from veins) carries the products of digestion to the liver, and arterial blood brings blood from the heart.

In vertebrates the liver, gallbladder, and pancreas are important glands of the digestive system. They are called associated glands since they do not lie within the digestive system, unlike the mucus- and enzyme-secreting glands that occur within the walls of the digestive tract. All three glands release chemicals into the duodenum.

Liver and gallbladder

All vertebrates have a liver. It is the largest gland and, after the skin, the largest organ of the body. The liver has several tasks. When an animal is an unborn fetus, the liver makes red blood cells. After birth, the liver destroys old red blood cells. One task the liver always performs is to detoxify (remove poisons from) the blood. Plant eaters, in particular, eat a lot of toxins; many plants store chemicals in their leaves and stems to defend themselves. The liver also produces bile, and stores carbohydrates, proteins, and fats, converting them

into other materials when needed. The liver contains a huge number of blood vessels, more so than nearly every other organ.

The gallbladder is a small organ that stores bile and releases it into the duodenum when needed. Bile emulisifes fats, breaking them into small globules that provide a greater surface area on which digestive chemicals can act. Most, but not all, vertebrates have a gallbladder. Lampreys, hagfish, most birds, and a few mammals do not have one.

The pancreas

The pancreas is an organ that makes and releases pancreatic juice, which largely comprises the protein-splitting enzyme trypsin. Other enzymes in the juice break down carbohydrates and fats. The pancreas also produces the hormones insulin and glucagon, which are proteins that regulate the levels of glucose (a sugar) in the blood. All vertebrates have a pancreas, though it is not always present as a single, discrete organ. The pancreas of lampreys and hagfish are gland-containing tissues spread throughout the intestinal submucosa and on the liver. Tetrapods (reptiles, amphibians, birds, and mammals) have a single discrete organ, as in humans.

Invertebrate equivalents

Invertebrates have neither a liver, a gallbladder, nor a pancreas. Some, however, have tissues that perform similar tasks. Lancelets are fishlike invertebrates that are close relatives of vertebrates. They have a pouch in their gut, the hepatic cecum, which develops in a similar position to the liver of a vertebrate. Veins bring blood carrying the products of digestion to the organ. Its tasks differ from those of a true liver, however, since it is where digestive enzymes are made and food is absorbed.

Earthworms have chlorogogen, a yellowish tissue circling the gut. It functions like a liver, storing glucose as glycogen and releasing it when needed and breaking down toxins. It also acts as a fat store and makes hemoglobin (the oxygen-carrying particle in the blood).

Excretory system

All life-forms must rid their body of toxins and wastes. In animals this is performed by a variety of excretory systems, including the kidneys in vertebrates. The kidneys produce urine and share ducts with the reproductive system. For that reason, both the urinary and reproductive systems are sometimes termed the urogenital system.

Excretion not only expels wastes. It also ensures that the body maintains the correct balance of vital chemicals—a regulatory function termed homeostasis. Aquatic life-forms, for example, balance the levels of salts in their body to prevent the excessive osmosis (diffusion) of water out of or into the body. Diffusion is the tendency for particles to move from a region of high concentration to a region of low concentration. Osmosis is the diffusion of water across a partly porous membrane, such as a cell membrane or nonwaterproof skin. Creatures that live in hot or dry places must balance salt levels to conserve water. Kidneys are vital water-saving organs for terrestrial vertebrates.

Defecation and urination

The digestive system voids feces at the anus or cloaca. Feces mostly comprises the undigested remains of food, along with other wastes such

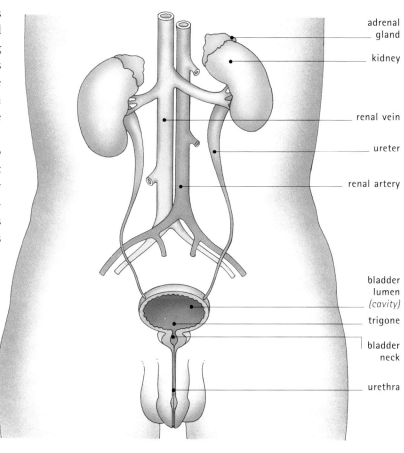

adrenal gland

kidney

renal vein

ureter

renal artery

bladder lumen (cavity)

trigone

bladder neck

urethra

▲ URINARY SYSTEM
Human male

In humans, other mammals, and many other vertebrates, the kidneys produce urine. The ureters, bladder, and urethra are vital for the passage, storage, and excretion of urine. The bladder is a hollow organ of varying capacity. It has a muscular coat that empties the organ when it contracts.

COMPARATIVE ANATOMY

Mammalian cloacae

A cloaca is a common chamber through which feces and urine pass and which also contains the genital opening. Cloacae occur in reptiles, birds, and amphibians, but are found in just a few types of mammals. These are the monotreme, or egg-laying mammals, which include the platypus and two species of echidnas.

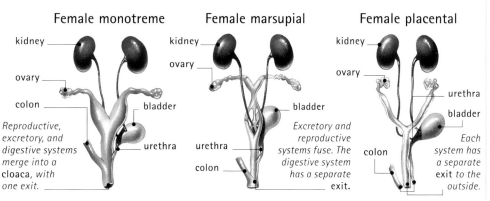

Female monotreme

kidney

ovary

colon

bladder

urethra

*Reproductive, excretory, and digestive systems merge into a **cloaca**, with one exit.*

Female marsupial

kidney

ovary

urethra

colon

bladder

exit.

Excretory and reproductive systems fuse. The digestive system has a separate exit.

Female placental

kidney

ovary

urethra

bladder

colon

Each system has a separate exit to the outside.

▶ **KIDNEY**

Mammal

The kidneys of birds and reptiles are similar to those of mammals. All have tiny coiled tubes called nephrons, which filter waste from the blood and reabsorb water and nutrients. In mammals the final result is urine, which is expelled via the bladder.

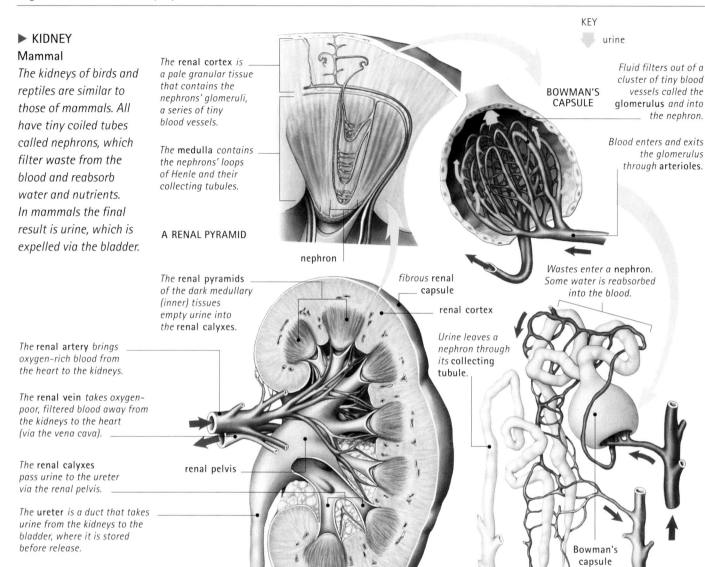

The renal cortex is a pale granular tissue that contains the nephrons' glomeruli, a series of tiny blood vessels.

The medulla contains the nephrons' loops of Henle and their collecting tubules.

A RENAL PYRAMID

nephron

The renal pyramids of the dark medullary (inner) tissues empty urine into the renal calyxes.

The renal artery brings oxygen-rich blood from the heart to the kidneys.

The renal vein takes oxygen-poor, filtered blood away from the kidneys to the heart (via the vena cava).

The renal calyxes pass urine to the ureter via the renal pelvis.

The ureter is a duct that takes urine from the kidneys to the bladder, where it is stored before release.

renal pelvis

LEFT KIDNEY CROSS SECTION

KEY

↓ urine

Fluid filters out of a cluster of tiny blood vessels called the glomerulus and into the nephron.

BOWMAN'S CAPSULE

Blood enters and exits the glomerulus through arterioles.

fibrous renal capsule

renal cortex

Wastes enter a nephron. Some water is reabsorbed into the blood.

Urine leaves a nephron through its collecting tubule.

Bowman's capsule

loop of Henle

NEPHRON

as the remnants of broken-down red blood cells. This process is called defecation. The urinary system excretes nitrogen-based wastes in a process called excretion. These wastes are made from the by-products of vital cellular processes collectively termed metabolism. Ammonia is a by-product of protein break down. This gas is toxic to most animals. In small aquatic animals it dissolves in water and diffuses out of the animal through the skin. Mammals convert ammonia into a chemical called urea, which dissolves into water and is passed out of the body in urine. Birds, reptiles, and terrestrial insects convert metabolic wastes into uric acid. This does not easily dissolve in water, so less water is needed to excrete it. Uric acid is excreted in the form of a paste.

Vertebrate excretion

Many vertebrates, including all mammals, have a pair of kidneys or kidneylike organs. The mammalian kidney regulates osmosis, balances chemicals, and produces urine. These functions are performed by separate organs in many non-mammals. The gills of many fish, for example, remove nitrogen compounds from the body. Many seabirds have glands that excrete salt, enabling them to drink seawater. Birds and reptiles do not have a bladder. Wastes are converted into uric acid. This is emptied into the end of the digestive tract and excreted via the cloaca with feces. Amphibians have a large bladder to store water in when they are on land. Also, the amphibian kidney increases or decreases its rate of filtration depending on whether the animal is in or out of the water and needs to conserve or expel water.

Amphibians and freshwater fish expel a lot of urine. This is because water constantly moves into the body through osmosis (since their body is saltier than the surroundings). Marine fish constantly lose water to their saltier surroundings. They must swallow a lot of seawater to counteract this, and conserve as much water form urine as possible. Most marine invertebrates are of similar saltiness to the seawater around them, so water neither moves in nor out.

Invertebrate excretion

There are four main types of invertebrate excretory organs—nephridia, renal glands, coxal glands, and Malpighian tubules. Mollusks have one or two kidneylike organs called renal glands, which filter metabolic wastes from body fluids. They are excreted as ammonia, ammonium chloride (in octopuses), or uric acid (in slugs and snails).

Aquatic arthropods such as crustaceans have paired coxal glands, which open at the bases of limbs or antennae. The gland is a coiled tube that empties urine into a bladder or a duct to the bladder. The tube begins as a small sac called the celomic sac, which filters the blood in a similar way to the glomerulus and Bowman's capsule of the vertebrate kidney.

Cellular elimination

Plants must protect themselves against toxins, too. Many plants contain poisons that deter animals from eating them. A plant cell typically contains a membrane-bound, fluid-filled space called a vacuole. Plants isolate poisons within their vacuoles. Coffee plants store caffeine within their vacuoles; tobacco plants store nicotine in theirs. Microorganisms expel poisons by enclosing them in vesicles. The vesicles merge with the cell membrane then empty their contents outside.

BRIDGET GILES

FURTHER READING AND RESEARCH

Arnold, Nick. 1999. *Horrible Science: Disgusting Digestion*. Scholastic: Danbury, CT.

Nephridia and Malpighian tubules

Earthworms have excretory organs called nephridia. Each nephridium is a long, fine tubule that opens in the body cavity and leads to the outside, dumping watery wastes onto the skin. Flatworms have excretory organs called protonephridia. They lead to clusters of cells called flame cells.

Insects use organs called Malpighian tubules to excrete wastes. Some insects have one pair, others have more than a hundred. Malpighian tubules start in the body cavity, where they are bathed in hemolymph (equivalent to blood). Wastes are pumped into the tubule from the blood. The tubules do not open onto the exterior but instead empty into the digestive tract, at the junction of the midgut and hindgut. The urine produced by the tubules passes through the rectum, where water and other useful chemicals are reabsorbed.

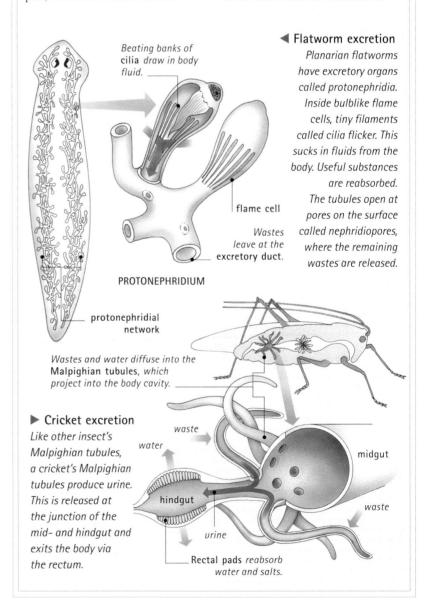

Beating banks of **cilia** *draw in body fluid.*

◀ Flatworm excretion
Planarian flatworms have excretory organs called protonephridia. Inside bulblike flame cells, tiny filaments called cilia flicker. This sucks in fluids from the body. Useful substances are reabsorbed. The tubules open at pores on the surface called nephridiopores, where the remaining wastes are released.

flame cell

Wastes leave at the **excretory duct.**

PROTONEPHRIDIUM

protonephridial network

Wastes and water diffuse into the **Malpighian tubules,** *which project into the body cavity.*

▶ Cricket excretion
Like other insect's Malpighian tubules, a cricket's Malpighian tubules produce urine. This is released at the junction of the mid- and hindgut and exits the body via the rectum.

waste

water

waste

midgut

hindgut

urine

waste

Rectal pads *reabsorb water and salts.*

Dolphin

ORDER: Cetacea SUBORDER: Odontoceti
FAMILY: Delphinidae GENUS: *Delphinus*

There are two species of common dolphins—a widely distributed short-beaked species that occurs in saltwaters from open ocean to inland seas, and a long-beaked species that favors coastal waters. Both have adapted to life in the warm surface waters. With sleek, streamlined bodies, they are fast-moving predators of fish and squid.

Anatomy and taxonomy
Scientists categorize all organisms into taxonomic groups based partly on anatomical features. The two species of common dolphins belong to the family of oceangoing dolphins and blackfish, the Delphinidae, which includes about 36 species in all. Oceanic dolphins, in turn, belong to the suborder Odontoceti, which includes river dolphins, porpoises, beaked whales, and sperm whales. There are about 73 species of toothed whales, although experts still occasionally discover new species.

The distinction between whales, dolphins, and porpoises is based on size and on anatomy. Strictly, all cetaceans are whales—whatever their size—but most people use the term *whale* for larger species only. The term *dolphin* refers to one family of small- to medium-sized oceangoing cetaceans, the Delphinidae; and four freshwater families, the river dolphins. Dolphins have cone-shaped teeth. The word *porpoise* refers to small cetaceans of the family Phocoenidae that have chisel- or spade-shaped teeth.

● **Animals** These organisms are multicellular and depend on other organisms for food. Animals differ from other multicellular life-forms in their ability to move from one place to another (in most cases, using muscles). They generally react rapidly to touch, light, and other stimuli.

● **Chordates** At some time in its life cycle a chordate has a stiff, dorsal (back) supporting rod called the notochord. It runs along most of the length of the body.

▶ *The common dolphin has only recently been split by biologists into two species, but short- and long-beaked versions are very difficult to tell apart. In waters around Australia both forms appear to be present. However, DNA analysis of this population has shown that all are actually short-beaked common dolphins. However, they are much more variable in size and shape than elsewhere in the world and some look just like the long-beaked form.*

Animals
KINGDOM Animalia

Chordates
PHYLUM Chordata

Vertebrates
SUBPHYLUM Vertebrata

Mammals
CLASS Mammalia

Placental mammals
SUBCLASS Eutheria

Cetaceans
(whales, dolphins, and porpoises)
ORDER Cetacea

Baleen whales
SUBORDER Mysticeti

Toothed whales
SUBORDER Odontoceti

Beaked whales
FAMILY Ziphiidae

Sperm whales
FAMILIES Physeteridae and Kogiidae

Narwhal and beluga
FAMILY Monodontidae

Oceanic dolphins and blackfish
FAMILY Delphinidae

Porpoises
FAMILY Phocoenidae

River dolphins
FAMILIES Iniidae, Lipotidae, Pontoporidae, Platanistidae

Long-beaked common dolphin
GENUS AND SPECIES *Delphinus capensis*

Short-beaked common dolphin
GENUS AND SPECIES *Delphinus delphis*

● **Vertebrates** The vertebrate notochord develops into a backbone made up of units called vertebrae. Vertebrate muscle blocks are generally bilaterally symmetrical about the skeletal axis—those one one side of the backbone are the mirror image of those on the other side.

● **Mammals** Mammals are warm-blooded vertebrates with hair. Females have mammary glands that produce milk to feed their young. Mammals have a single lower jawbone that hinges directly to the skull, and their red blood cells do not contain nuclei.

● **Placental mammals** Placental mammals nourish their unborn young through a placenta, a temporary organ that forms in the mother's uterus during pregnancy.

● **Cetaceans** These mammals are supremely adapted for life in water, where they spend their entire lives. Cetaceans have bodies that are streamlined to minimize drag as they swim through the water. Cetacean nostrils have moved over millions of years of evolution from the front of the head to the top. This enables breathing at the sea surface; the nostrils end at one or two blowholes. Cetaceans have paddlelike tails. Their forelimbs form flippers and lack visible digits. Cetaceans do not have functional hind limbs. Most have a dorsal (back) fin that aids steering and provides stability against rolling from side to side when swimming.

● **Baleen whales** There are 12 species of baleen whales, including most of the larger whales. Instead of teeth, baleen whales have fringes called baleen plates, which hang from the upper jaws and strain small fish or shrimplike crustaceans from the water. All baleen whales have two blowholes, side by side.

▲ *Dolphins have beaklike snouts containing pointed, conelike teeth that are ideal for holding slippery fish.*

● **Toothed whales** These whales have teeth rather than baleen. In most of the 73 or so species, the jaws are extended into a beaklike snout. The forehead bulges upward, enclosing a "melon," a fatty structure that focuses sound waves to enable the whale to echolocate (create an image of the surroundings using sound). All toothed whales have a single blowhole.

● **Sperm whales** The three species of sperm whales have a huge, square head. These deep divers have a wax-filled structure in the head called the spermaceti organ. This focuses echolocation sounds and helps adjust buoyancy

FEATURED SYSTEMS

EXTERNAL ANATOMY Dolphins are toothed whales with a sleek, streamlined body, flippers, and a powerful, horizontally flattened tail for swimming. *See pages 191–193.*

SKELETAL SYSTEM The backbone acts as an anchor for muscles that flex the body and fins and that raise the tail up and down. *See pages 194–196.*

MUSCULAR SYSTEM Large muscles power vertical movements of the tail; a system of fibers, acting as springs, stores and releases energy, making tail movement an efficient process. *See pages 197–198.*

NERVOUS SYSTEM The dolphin brain is relatively complex, with a large and highly folded cerebral cortex. This reflects the animal's social behavior; its life in a complex, three-dimensional habitat; and the high processing power needed to interpret sounds created for echolocation. It is also a sign of intelligence. *See pages 199–201.*

CIRCULATORY AND RESPIRATORY SYSTEMS These systems ensure that oxygen reaches vital organs during long dives, while minimizing the dangers of dissolved nitrogen at high pressures. *See pages 202–203.*

DIGESTIVE AND EXCRETORY SYSTEMS Dolphins and other toothed whales swallow their prey whole or in large chunks. The first part of the complex stomach breaks down food mechanically. *See pages 204–205.*

REPRODUCTIVE SYSTEM Reproductive adaptations for life underwater include internal male sex organs, hidden mammary glands, and giving birth tail first. *See pages 206–207.*

▲ *As with most oceanic dolphins, this bottle-nosed dolphin is countershaded—dark above, lighter below. This camouflages the animal against prey and predators from both above and below.*

● **Narwhal and beluga** These two species are medium-sized whales that live in arctic and subarctic waters and feed on fish, squid, and crustaceans such as crabs and shrimp. The beluga is white or pale yellow. The narwhal has a dark mottled back and a pale underside. Male narwhals sport an impressive unicorn-like tusk. Both species have a flexible neck. They are the only whales that can dramatically alter their facial expressions. Belugas and narwhals gather in large numbers at breeding time.

● **Beaked whales** The 21 or so species of deep-diving beaked whales have pointed snouts that contain few teeth. These animals feed on squid. The males of most beaked whales have two or four teeth in the lower jaw and none in the upper. The teeth probably serve as weapons in fights between males. Most females have no erupted teeth at all.

● **Porpoises** The six species of blunt-headed porpoises are mainly coastal but may occur in rivers or the open sea. They have spade- or chisel-shaped teeth for grasping prey.

● **River dolphins** Four of the five species of river dolphins live in large, muddy rivers of South America or Asia; the fifth species lives in South American coastal waters. River dolphins have a long narrow beak, a highly domed forehead, and tiny eyes. In the cloudy waters in which they live, eyesight is almost useless. They rely on sophisticated echolocation to find their way about and to detect prey.

● **Killer whales and pilot whales** Six species of whales are called blackfish because of their dark color. They belong to the Delphinidae, the same family as the oceanic dolphins, but blackfish are larger than dolphins and have large flippers, a blunt head lacking a beak, and fewer teeth. Blackfish are hunters that often work together to catch fish and squid. Killer whales regularly kill and eat other cetaceans, and may even attack giants such as blue whales.

● **Oceanic dolphins** Of the 30 or so species of oceanic dolphins, about half (including the common and bottle-nosed dolphins) have a prominent beak. Almost all oceanic dolphins have more than 100 small, conical teeth for grasping fish and squid.

● **Common dolphins** The long-beaked common dolphin has a pronounced beak and a slightly longer body and head than the short-beaked form, and it is less boldly colored. Both species are social, often traveling in groups of more than 100. Long-beaked dolphins live around coasts; short-beaked dolphins live in deeper waters.

COMPARATIVE ANATOMY

Dorsal fins

Killer whales have very tall dorsal fins. Most dolphins have more moderately sized ones, while the sperm whale has an irregular ridge or hump and the narwhal, beluga, and finless porpoise have no dorsal fin at all. In many cases, the dorsal fin serves as a keel that helps prevent the whale from rolling (rotating to one side) as it swims. In some species this fin may also serve as a temperature regulation device; by holding the dorsal fin above the water, the animal can lose or absorb heat. Male killer whales have larger dorsal fins than the females, and this feature may help individuals identify the sex and status of others.

External anatomy

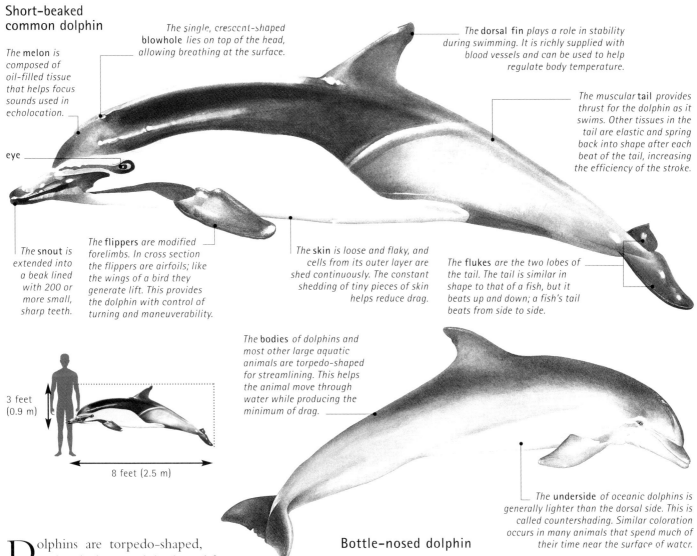

Short-beaked common dolphin

The **melon** *is composed of oil-filled tissue that helps focus sounds used in echolocation.*

The single, crescent-shaped **blowhole** *lies on top of the head, allowing breathing at the surface.*

The **dorsal fin** *plays a role in stability during swimming. It is richly supplied with blood vessels and can be used to help regulate body temperature.*

eye

The muscular **tail** *provides thrust for the dolphin as it swims. Other tissues in the tail are elastic and spring back into shape after each beat of the tail, increasing the efficiency of the stroke.*

The **snout** *is extended into a beak lined with 200 or more small, sharp teeth.*

The **flippers** *are modified forelimbs. In cross section the flippers are airfoils; like the wings of a bird they generate lift. This provides the dolphin with control of turning and maneuverability.*

The **skin** *is loose and flaky, and cells from its outer layer are shed continuously. The constant shedding of tiny pieces of skin helps reduce drag.*

The **flukes** *are the two lobes of the tail. The tail is similar in shape to that of a fish, but it beats up and down; a fish's tail beats from side to side.*

The **bodies** *of dolphins and most other large aquatic animals are torpedo-shaped for streamlining. This helps the animal move through water while producing the minimum of drag.*

3 feet (0.9 m)

8 feet (2.5 m)

Bottle-nosed dolphin

The **underside** *of oceanic dolphins is generally lighter than the dorsal side. This is called countershading. Similar coloration occurs in many animals that spend much of their time near the surface of water.*

Dolphins are torpedo-shaped, with a sleek, smooth body and few surface projections. The body is streamlined to reduce drag (the resistance of water to movement through it), enabling the animal to swim rapidly and efficiently.

Whales and dolphins evolved from land-living ancestors. Over the course of millions of years, cetaceans have lost many land mammal features, such as the hind limbs and external ears. Other features have been modified for an aquatic lifestyle, such as the forelimbs, which have become flippers.

Smooth skins

Cetacean skin is smooth and almost hairless. Instead of relying on hair to provide heat insulation, whales and dolphins have a thick layer of fat, or blubber, beneath the skin. This helps minimize heat loss to the surrounding water. Blubber serves as a reserve of fat, and its elasticity (stretchiness) makes each tail beat more efficient. The skin and blubber are attached to each other by a network of small projections called dermal papillae. These form ridges on the skin similar to those on human fingers that form fingerprints. The ridges guide water over the dolphin's skin. This promotes a smooth, laminar (layered) flow of water. A more turbulent flow with swirls and eddies would cause increased drag. The skin lacks sweat glands but releases an oily substance that helps the animal slide effortlessly through the water.

▲ **Short-beaked common dolphin and bottle-nosed dolphin**
The skin of these cetaceans is both smooth and hairless, and it flakes away almost constantly. These drag-reducing measures help the animals save energy as they swim swiftly through the ocean.

Blubber

Like many other warm-blooded marine animals, such as seals, sea lions, walruses, and sea cows, whales have a thick layer of fat called blubber beneath the skin. Apart from insulating the animal against the cold, blubber is a food store and provides buoyancy.

▼ *A cross section through dolphin skin.*

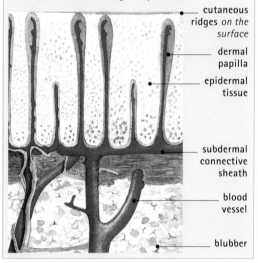

cutaneous ridges *on the surface*

dermal papilla

epidermal tissue

subdermal connective sheath

blood vessel

blubber

Limb morphology

Cetacean front limbs form flippers used for steering. The hind limb bones are now absent, and the connecting pelvic bones are vestigial (greatly reduced in size and no longer serving their original purpose). The tail is broadened

horizontally into two blades, or flukes. Up-and-down movement of the tail powers swimming; it enables most species to leap clear of the water, during porpoising in dolphins or breaching in whales.

Toothed whales breathe through a single blowhole on top of the head. Compared with the heads of other mammals, a dolphin's head is elongated and grades into the trunk with no visible neck or shoulders. Most types of dolphins, including common and bottle-nosed dolphins, have a prominent snout, or beak.

▲ *When swimming at top speed, dolphins leap from the water. This is called porpoising. It allows them to take a breath without needing to slow down at the water's surface.*

Ancient forerunners of the whales

Whales and dolphins descend from condylarths, an ancient group of ungulates (hoofed mammals). Modern artiodactyl (even-toed) ungulates are whales' closest relatives. They include animals like hippopotamuses and antelope. The oldest fossil cetacean discovered to date is the 52-million-year-old *Pakicetus*, which belonged to a group called the Archaeoceti, or ancient whales. *Pakicetus* is known only from its skull; features associated with hearing show that this wolf-sized animal lived at least partly on land. Within a few million years the archaeocete whales were fully aquatic.

Rodhocetus, for example, had well-developed hind limbs but its pelvic vertebrae (unlike those of land mammals), were not fused to form a rigid sacrum. Over the next 10 million years, the archaeocetes diversified to include some true giants, such as *Basilosaurus*, which measured up to 60 feet (18 m) long. This beast still had complete hind limbs, including knee joints and toes. By this stage, though, these tiny limbs were useless for propulsion through the water. Around 33 million years ago, the Archaeoceti split into the two main modern whale groups: toothed and baleen whales.

Arthropod foreguts and midguts

The foreguts of arthropods, including insects and spiders, also have a chitinous lining. Arthropods have an external skeleton called an exoskeleton. The exoskeleton's outer layer is the chitinous cuticle, which extends into and lines the foregut and hindgut (intestines). The midgut is not lined by cuticle.

Backward-pointing spines or "teeth" might project from the surface of an insect's foregut. An insect's muscular pharynx sucks and swallows, and the esophagus may be enlarged to form a crop. The crop empties into the "stomach," or mesenteron (midgut). Some insects have a muscular gizzard between crop and midgut. Invertebrate gizzards generally contain teeth or grinding plates rather than gastroliths, tiny stones that occur in the gizzards of birds and their ancestors, dinosaurs. Digestion and absorption of the products of the process occur in the insect midgut. The cells that line the midgut have folded surfaces that form fingerlike projections called villi. These are covered by even tinier bumps called microvilli. The projections increase the surface area across which the products of digestion can be absorbed. Midgut pouches called ceca may further increase the surface area.

A crustacean foregut might be a simple tube. However, the foreguts of crabs, lobsters, and other decapods (10-legged forms) contain a structure called the gastric mill. This is a series of hard plates that grind food. Hairlike setae prevent large particles of food from entering the midgut. The midgut often contains one or more diverticula, or pouches, inside which digestion occurs.

Mollusk foreguts and stomachs

Most filter-feeding bivalve mollusks, such as clams and oysters, have gills that pick up oxygen and obtain particles of food from the water. The complex stomach contains one end of a rodlike crystalline style, which rotates and grinds against a hard area of the stomach's wall. As it rotates it dissolves, releasing enzymes that begin to digest the filtered food. In snails and other nonbivalves the foregut includes glands and "teeth" made from tough chitin. These teeth, called the radula, bite, tear, and scrape food materials.

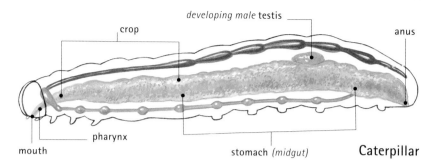

developing male testis

crop

anus

pharynx

mouth

stomach (midgut)

Caterpillar

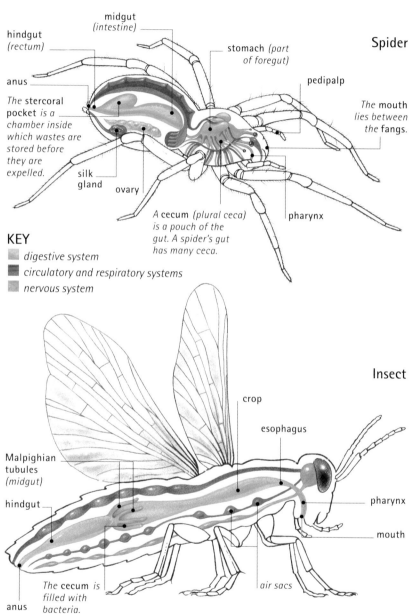

hindgut (rectum)

midgut (intestine)

stomach (part of foregut)

anus

pedipalp

The stercoral pocket is a chamber inside which wastes are stored before they are expelled.

The mouth lies between the fangs.

silk gland

ovary

pharynx

A cecum (plural ceca) is a pouch of the gut. A spider's gut has many ceca.

Spider

KEY

- digestive system
- circulatory and respiratory systems
- nervous system

Insect

crop

esophagus

Malpighian tubules (midgut)

hindgut

pharynx

mouth

anus

The cecum is filled with bacteria.

air sacs

▲ DIGESTIVE SYSTEMS
Arthropods

Caterpillars are eating machines, and their large crop, stomach, and intestines reflect this. The pharynx and stomach of a spider are attached to strong muscles. When the muscles contract, the pharynx and stomach dilate. This creates a vacuum that sucks in the spider's liquid food. The ceca of a plant-eating insect might be filled with microorganisms that help digest tough cellulose.

179

Whale facial expressions

Of the cetaceans, only the beluga whale and narwhal can alter their facial expressions. The beluga's neck and mouth are highly flexible. It also has a very large melon, the forehead bulge containing tissues that focus echolocation signals.

Belugas communicate by sound, producing a wide range of clicks, cheeps, and bell-like tones. By altering the shape of the mouth, lips, and melon, beluga whales appear to smile, frown, and whistle. The meaning of these expressions is unknown.

Smiling, pursed lips (left) and a grumpy frown (right) have specific but unknown meanings in beluga communication.

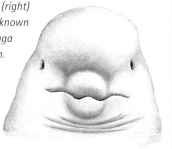

pattern on their flanks that is created by combinations of yellow, white, gray, and black. Bold coloration serves a number of functions.

Cetaceans echolocate and communicate with each other using sound, but visual identification is important at close range. Individuals can probably recognize the age, sex, and status of of other individuals by sight as well as sound. Patches of color may also act as movement and orientation signals. As a common dolphin banks to change direction, new colors and shapes are brought into sharper view. Other individuals respond, helping a school of dolphins turn as one and move in a tight formation when hunting prey or avoiding predators. Contrasting colors also help break up the body outline, which can confuse attacking predators or fleeing prey.

▼ *A beluga whale has a very large melon, which plays a vital focusing role during echolocation and is also used to convey visual signals to other belugas.*

Dolphins and whales do not have external ears as most other mammals do. Two tiny openings lead from the outside directly to the hearing organs, but they are largely non-functional. Instead, toothed whales hear by channeling sound waves through the jaw to the inner ears. Most dolphins have forward-looking eyes and can see well both above and below the water. River dolphins' eyes are not so good. These dolphins rely on echolocation to find their way around in murky waters.

The dolphin's external reproductive organs are tucked away inside the body. In this way, streamlining of the body is not compromised. The male's penis is hidden behind muscular folds and emerges only when the male is aroused and ready for copulation. The teats of the female's mammary glands are housed within slits and appear only for suckling.

Body color

Most whales are drab combinations of gray, black, brown, or white. However, some dolphins, particularly oceanic dolphins that congregate in large numbers, are more boldly marked. Long-beaked common dolphins, for example, have a characteristic "hourglass"

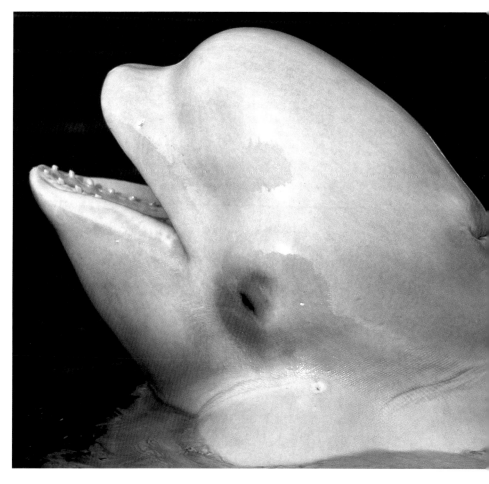

Skeletal system

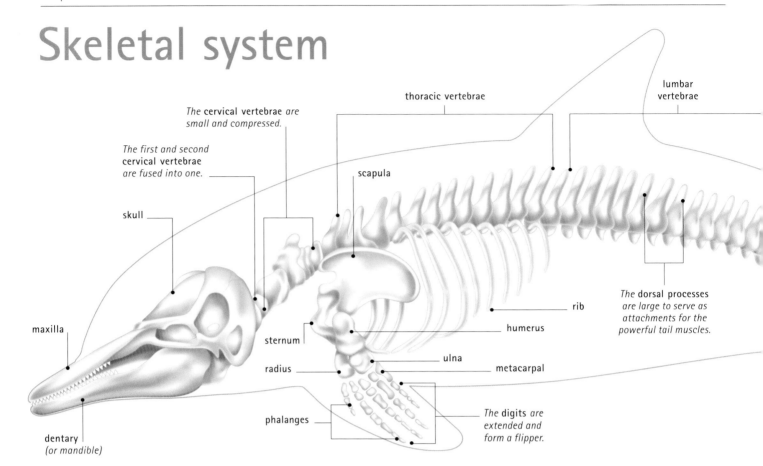

The cervical vertebrae are small and compressed.

The first and second cervical vertebrae are fused into one.

skull

maxilla

dentary (or mandible)

thoracic vertebrae

lumbar vertebrae

scapula

The dorsal processes are large to serve as attachments for the powerful tail muscles.

rib

humerus

sternum

ulna

metacarpal

radius

phalanges

The digits are extended and form a flipper.

▲ **Short-beaked common dolphin**
The bones of a short-beaked common dolphin. The remnants of the pelvic bones (not shown) lie below the junction of lumbar and caudal vertebrae.

In all vertebrates, the skeleton has four principal functions: it shapes and supports the animal's body; it protects vital internal organs such as the brain, heart, and lungs; it allows movement of body parts such as the head and limbs; and it enables locomotion, the movement of the whole animal.

A dolphin's skeleton is very different from that of a land mammal. Air does not provide physical support for an animal. Water is a relatively thick fluid that provides support to a swimming creature. Land-living mammals have a strong skeleton with limbs that raise the body off the ground. The legs act as vertical compression struts, rather like the piles that support a bridge. Mammals that spend their entire lives in water do not need this kind of support. In fact, a dolphin with its lungs full of air is effectively weightless in water. However, a cetacean's skeleton and muscles cannot support the animal out of water. If a large whale gets stranded on the shore, it lies helpless and eventually suffocates under its own weight.

Skull and jaws

In any vertebrate, the skull's primary function is to protect the brain. The mammalian skull, however, is much larger than required for this function alone. It also contains large air spaces that connect to the nasal passages. When air is breathed in through the nose, its winding route through the skull ensures that the air is warmed and moistened before it reaches the lungs. This keeps heat loss in the lungs to a minimum and helps stop the membranes of the trachea (windpipe) from drying out.

CLOSE-UP

Light bones

The bones of whales are astonishingly light. Unlike those of land mammals, they do not need to be strong enough to support the animal in air. A whale's long bones consist of a thin outer shell of hard, compact bone, while the inside contains thin bony bars with large spaces in between. This is called spongy bone.

Whale skeletons contain far more spongy bone than those of land mammals. In adult whales, spongy bone is filled with fatty, yellow marrow. It also contains red bone marrow, which produces red and white blood cells, and platelets. In whales, the marrow-filled spongy bone also contributes to the animal's buoyancy.

The long jaws can sweep from side to side to catch fish.

Pointed teeth hold prey once it is caught.

Boutu (Amazonian river dolphin)

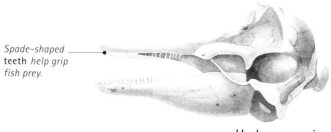

Spade-shaped teeth help grip fish prey.

Harbor porpoise

Pointed, deeply anchored teeth for holding struggling prey.

Robust jaws can withstand the impact of large prey.

Killer whale

As in other mammals, the sacral vertebrae lie between the lumbar and caudal vertebrae and fuse with the pelvis. In dolphins, the sacral vertebrae are indistinguishable from the lumbar vertebrae and are usually counted as lumbar vertebrae.

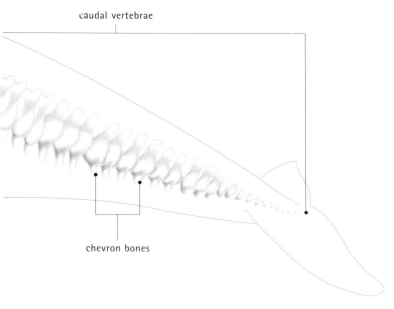

caudal vertebrae

chevron bones

In comparison with that of a zebra, the skull of a toothed whale is elongated (stretched lengthwise). This produces a jaw arrangement suitable for grasping narrow prey such as fish. The beak of most dolphins is streamlined, enabling it to cut through the water when swimming. A depression at the front and top of the skull accommodates the fat-containing melon, which is important for echolocation. The dolphin breathes through a blowhole at the top of the skull, so only a small part of the body needs to break the water's surface.

The backbone

In land mammals, the backbone acts as a firm girder that supports the animal's weight. In whales, water supports the animal's weight. The backbone of a whale has become more important for the attachment of muscles that propel the animal through water. In most toothed whales, muscles account for about 40 percent of the animal's weight.

The trunk region of a whale is fairly rigid. The head and neck region can bend up and down to some extent, and the caudal (tail) region is very flexible. The varying degrees of movement are reflected in the number of vertebrae in different regions of the spine—more vertebrae generally allow for greater flexibility. The extent of firm or flexible connecting tissues between the vertebrae is also of importance.

Common dolphins have 75 vertebrae—more than most other mammals. Extra vertebrae in the tail provide the muscle attachments and flexibility required for swimming. Whales do have a neck, but in most species it is short and rigid. Dolpins can nod their head, but not swivel it or turn it to look backward. The limited movement of the neck is caused by the fusion of the first and second cervical vertebrae (the atlas and axis). The bottle-nosed dolphin is an exception, since its

▲ SKULLS
Boutu and harbor porpoise are specialist fish eaters, while the killer whale eats much larger (often mammalian) prey. These skulls are not shown to scale.

◄ TEETH
Dolphins and porpoises have very different teeth. Those of dolphins are long and pointed, while porpoise teeth are shovel shaped. Both groups, however, feed mainly on fish and squid. Some porpoises have horny bumps on the gums between the teeth. These help grip slippery prey, although they are worn down in older porpoises.

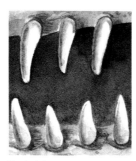

Dolphin

Porpoise

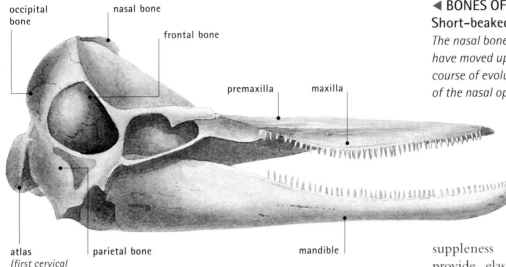

occipital bone

nasal bone

frontal bone

premaxilla

maxilla

atlas
(first cervical vertebra)

parietal bone

mandible

◄ BONES OF THE SKULL
Short-beaked common dolphin
The nasal bones of dolphins and other cetaceans have moved up high onto the forehead over the course of evolution. This enabled the movement of the nasal opening to form the blowhole.

▲ **WHALE TOOTH**
This is a cross section through a killer whale's tooth. Each line represents a year's growth. By cross-sectioning the teeth of a stranded cetacean, biologists can get a good idea of how long the animal lived.

neck can bend up or down by 45 degrees. The five other neck vertebrae are compacted together in most whales, including all the oceanic dolphins. The short, rigid neck helps keep the head from wobbling from side to side when the animal is swimming. Head wobbling would cause drag and a sideways movement called yaw.

Most of the 13 or 14 thoracic vertebrae of common dolphins are fairly rigid. In whales only the first few are shaped to allow much flexibility and movement. Most vertebrae have flat faces that lie against each other and therefore allow little movement.

The lumbar (lower back) region of the spine is extraordinarily long in whales. The individual lumbar vertebrae are large, with broad surfaces and large extensions

for the powerful tail muscles to attach to. Dolphins have at least 30 caudal (tail) vertebrae. These bones have cushioning pads of cartilage that give great suppleness to the tail. The pads also provide elasticity. The joints between the vertebrae cushion the compressive (squeezing) forces that occur when the tail muscles contract. Most of the caudal vertebrae have bony extensions on top and below for muscle attachment. The last few caudal vertebrae are simple, flattened bones. They provide the central support for the broad tail flukes.

Limbs and their supports

In land-living mammals, forelimbs and hind limbs are connected to the spine by the limb girdles. The fore, or pectoral, girdle contains two scapulae (shoulder blades) that allow most land mammals to move their forelimbs flexibly through wide arcs.

The hind, or pelvic, girdle is more robust than the pectoral girdle, and is anchored to the backbone through sacral vertebrae that are themselves fused together. Over millions of years, the front legs of whales' increasingly aquatic ancestors evolved to become flippers. The humerus (upper forelimb bone) of a typical whale flipper connects via a joint (the "elbow" in humans) to the ulna and radius—these are the lower forelimb bones. The basic setup is similar to a human arm. However, a dolphin's upper and lower forelimb bones are much shorter than a human's, while the digits are considerably longer. The second and third digits are very long, with extra phalanges providing support. The whole arrangement is enclosed by skin and connective tissue so the outline of the limb is smooth and streamlined. Cetacean hind limbs have disappeared. Of the pelvic girdle, just a few tiny bones remain. Biologists refer to these as vestigial structures.

CLOSE-UP

Losing legs

During the evolutionary transition from life on land to an aquatic existence, whales' hind limbs shrunk in size; today's whales have entirely lost their hind limb bones. Without hind limbs to support, the pelvic bones have also become greatly reduced. Modern whales have a small bone that represent the remnants of the pelvis. This is usually functionless, although it is used to anchor the muscles of the penis in the males of some species.

Muscular system

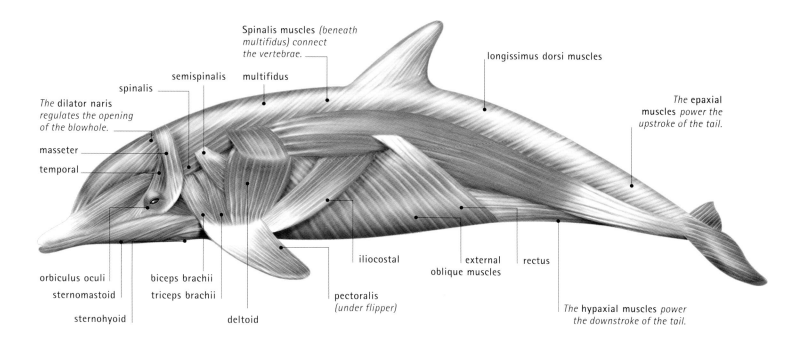

Spinalis muscles *(beneath multifidus)* connect the vertebrae.

longissimus dorsi muscles

semispinalis   multifidus

spinalis

The dilator naris *regulates the opening of the blowhole.*

The epaxial muscles *power the upstroke of the tail.*

masseter

temporal

orbiculus oculi
sternomastoid
sternohyoid

biceps brachii
triceps brachii

deltoid

iliocostal

pectoralis
(under flipper)

external
oblique muscles

rectus

The hypaxial muscles *power the downstroke of the tail.*

▲ **Short-beaked common dolphin**
The musculature of a short-beaked common dolphin.

As in other vertebrates, the movement of the bones of the skeleton is brought about by antagonistic pairs of muscles—sets of muscles that work against each other. The contraction of one set moves a bone in one direction, and the contraction of another set moves it in the opposite direction.

The flippers of a dolphin are supported by bones within, but the dorsal fin and the tail flukes are not. In the case of the dolphin's tail, muscles running from the lumbar (lower back) and caudal regions of the backbone connect through tendons to systems of tough fibers in the flukes. Contraction of the muscles not only bends the tail as a whole, but also changes the orientation of the flukes. At certain stages of the power and recovery strokes of the swimming cycle, the tail flukes move through the water with ease; at other stages, they push against the water to propel the animal forward.

CLOSE-UP

Muscular oxygen stores

Working muscles have a high demand for oxygen. To gain energy to contract, they need oxygen to release energy through a series of chemical reactions—this is called aerobic respiration. Deep-diving sperm whales can stay underwater for up to two hours. How can the whale's muscles gain the oxygen they need when the whale is cut off from the supply of air at the surface?

The answer lies in the muscles themselves. They have their own stores of oxygen bound to a pigment protein called myoglobin. The myoglobin gradually releases its oxygen store during the whale's dive. Weight for weight, sperm whale muscle can store twice as much oxygen as that of any land mammal. Sperm whale muscle is dark in color because of the large quantities of myoglobin it contains.

The beating of the whale's tail is brought about by two main sets of muscles. One set, the epaxial muscles, lies above the backbone. The other set, the hypaxial muscles, lies below it. The epaxial muscles are much larger than the hypaxial muscles. Contraction of the epaxial muscles raises the tail. This is called the power stroke, since it provides most of a dolphin's thrust. Large epaxial muscles enable a strong power stroke.

Connecting muscles

A spiral of connective tissue called the subdermal sheath winds around the epaxial and hypaxial muscles. When the epaxial muscles contract to power the tail's upstroke, the hypaxial sheath stores energy like a coiled spring. The energy is released to help power the downstroke. This makes the dolphin's tail more powerful and energy-efficient than the size of its muscles would suggest. When the hypaxial muscles contract for the tail's downstroke, the epaxial sheath stretches in readiness for the upstroke.

Blubber helps, too. The blubber above and below the tail muscles is arranged in wedges that contain springlike collagen fibers. When the tail rises, the topmost wedge compresses, storing energy, which is released when the tail descends. Then the lower wedge compresses, storing energy for the upward rebound. Together with the dolphin's skin properties, this helps explain why dolphins can swim so fast—at 20 mph (32 km/h) or more.

▲ *A short-beaked common dolphin leaps from the ocean surface. All the power required to propel the animal into the air is generated by the tail.*

CLOSE-UP

Dolphin flukes and vortices

If a dolphin's tail simply beat up and down in the water, the animal could not move forward through the water. The flukes in the dolphin's tail must bend and alter their angle of attack at different points of the swimming stroke. Water leaves the trailing edge of the flukes in the form of swirls called vortices. Stages in the power and recovery stroke of a dolphin's tail include these:
1. Epaxial muscles contract, and the tail moves upward. This power stroke creates a vortex and a point of high pressure beneath the tail.

2. The dolphin moves forward and down.
3. At the top of the power stroke, the vortex leaves the trailing edge of the flukes.
4. The dolphin glides through the water with the tail contracted to offer minimum resistance.
5. The hypaxial muscles contract and slowly lower the tail in readiness for the next power stroke. This is called the recovery stroke; it is partly powered by elastic recoil of the tissues.
6. At the end of the recovery stroke the fluke is extended, ready for a new stroke cycle.

Nervous system

CONNECTIONS

COMPARE the echolocation system of a dolphin, which operates in water, with that of a *FRUIT BAT*, which functions in air.

COMPARE the size and shape of a dolphin's brain with those of other intelligent mammals, such as the *HUMAN* and the *CHIMPANZEE*.

▼ **Short–beaked common dolphin**
Important parts of the dolphin nervous system. Relative to body size, the dolphin brain is among the largest in the animal kingdom.

Dolphins and other toothed whales have a nervous system that is responsible for rapidly coordinating activities within the body and for responding to environmental changes outside the body. Coordination is brought about by electrical impulses that travel along nerve cells called neurons. They are bundled together to form nerves.

A vertebrate's nervous system is divided into the central nervous system (CNS), which consists of the brain and spinal cord, and the peripheral nervous system, or PNS. The PNS contains the nerves that connect the CNS with the sensory organs, such as the eyes and ears, and also with responding structures (effectors), such as muscles and glands.

Stimuli from the external environment, such as light waves or sound vibrations, stimulate receptors in sensory organs such as the eyes or inner ears. Sensory neurons deliver nervous impulses from these organs through the PNS to the CNS, where the brain or spinal cord processes them. The CNS then executes an appropriate response by sending nerve impulses along motor neurons to muscles—for example, to change behavior in response to the environmental stimuli.

The structure of the brain

Toothed whales, like mammals in general, have a brain with three main regions: the forebrain, the midbrain, and the hindbrain. The hindbrain contains the medulla and cerebellum. The medulla regulates automatic activities; it is involved in the control of physiological features such as body temperature, heartbeat, and breathing, for example. The cerebellum automatically controls body movements, making them smooth and coordinated. This structure is large in dolphins; it needs to be large to help control the animal's rapid, highly coordinated swimming. The midbrain, which contains the thalamus, is a relay center for directing nerve impulses from sensory organs to different parts of the forebrain.

The forebrain contains the cerebrum. This has an outer layer, the cerebral cortex, that is made up of gray brain matter. Gray matter contains numerous connections between nerve cells. The cerebral cortex is folded so it looks like the surface of a walnut. Folding increases the surface area and hence the number of connections. As a rule, the greater the degree of folding, the greater the

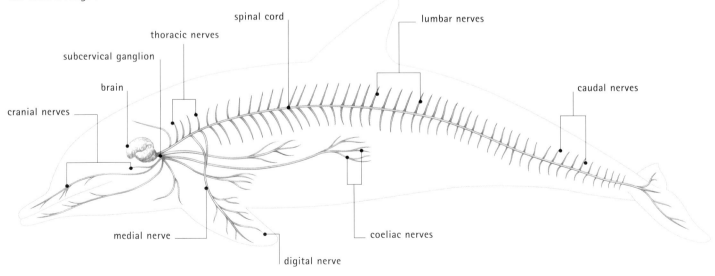

spinal cord

thoracic nerves

lumbar nerves

subcervical ganglion

brain

caudal nerves

cranial nerves

medial nerve

coeliac nerves

digital nerve

Stunning with sound?

Scientists have recorded wild dolphins hunting using loud, narrow beams of sound as they focus tightly on individual fish. These beams of sound may serve to stun or disorientate the dolphin's prey. This hypothesis is difficult for biologists to confirm because dolphins do not make such sounds in captivity. Some other cetaceans, such as group-feeding humpback whales, use loud sounds to confuse and terrify schools of fish. Humpbacks also blow out curtains of air bubbles to herd the fish into one place.

▶ **Sound hunters**
This pair of spotted dolphins may use beams of sound to stun their fish prey.

intelligence of the animal. In humans and toothed whales the cerebral cortex is highly folded. The cerebrum is responsible for memory, learning, and reasoning, and for processing complex sensory information, such as that relayed from the eyes and ears.

Touch, taste, and smell
Toothed whales such as dolphins have all the senses generally found in land-living mammals, although the sense of smell (olfaction) may be limited. Researchers have found taste receptors on the tongue of various species of toothed whales, and these may help the animals select food items. Toothed whales might also be able to taste chemicals released by other individuals that give clues to their availability for mating. Tasting the chemical signature of a current might also help whales navigate the oceans.

The skin of toothed whales is sensitive, especially the region around the blowhole. Touch is important in the social life of toothed whales. Stroking and touching are part of the courtship rituals in many species, and mothers and their calves touch regularly. This helps maintain their close bond.

Sight
Dolphins see well in both air and water. When hunting for flying fish, bottle-nosed dolphins are able to track their quarry as they glide through the air. Killer whales often spyhop when they are hunting; they raise their heads up out of the water and look

◀ *Dolphins produce high-frequency sounds to echolocate. Some overlap with the range of human hearing but most are many times higher.*

200,000 Hz

150,000 Hz

dolphin sonar

100,000 Hz

50,000 Hz

20,000 Hz

human hearing range

10,000 Hz
5,000 Hz
20 Hz

▼ *How dolphins make and receive echolocation signals.*

around for seals resting on ice floes. Light is more strongly refracted (bent) by the eye in air than it is in water. An eye that can see well in air cannot focus as sharply underwater. Whales get around this problem by using different parts of the eye for focusing in air and in water.

In most toothed whales, the field of vision of one eye overlaps that of the other. So, there is a large area toward the front where the vision is binocular (both eyes can focus on an object). This allows the animal to see well in three dimensions and to accurately judge the distance to prey and other objects. The eyes of toothed whales with large blunt heads, such as sperm whales, are set far back, and the animal has little or no binocular vision. Sperm whales do much of their hunting in or beyond the twilight zone, at depths beyond 3,300 feet (1,000 m). Sperm whales have less need for good vision

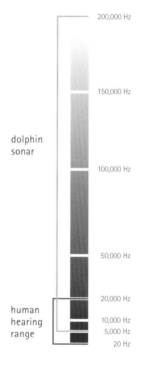

1. Ultrasonic clicks are produced by nasal sacs.

skull

2. Sounds are focused by the melon.

3. An echolocation pulse is emitted.

4. Echoes bounce back from objects.

5. Echoes are channeled along the **lower jaw**.

6. Sound is detected by **inner ear bones**, which are insulated from the skull by the **auditory bulla**.

inner ear bones

since little light reaches this part of the ocean. Vision is also less important for dolphins that live in murky water, such as river dolphins.

Hearing and echolocation

Sound travels much farther and faster in water than in air, and all toothed whales have good hearing. They can detect sounds that are higher pitched (ultrasonic) and lower pitched (infrasonic) than humans can hear. Several features make dolphin hearing very different to that of land mammals.

In toothed whales, a sound-conducting channel of fat in the lower jaw carries sound waves from the surrounding water to the bones of the middle ear; they pass the vibrations to the inner ear, where they are converted into nervous impulses that travel on to the brain. Only the lowest-pitched sounds travel through the dolphin's external ear—the channel normally used in land mammals. It is tiny in dolphins. Dolphins' hearing is extremely sensitive. It is aided by foam-filled spaces that cushion the inner ear from the skull, ensuring that other sounds do not interfere with those channeled out through the melon and those returning through the jaw. Toothed whales communicate with a wide range of sounds, including croaks, whistles, and squeaks. The beluga is sometimes called the "sea canary" for its melodious calls. Orcas, or killer whales, living in different localities have distinctive "dialects."

Toothed whales can hunt for food at night, or in deep or murky waters where there is little or no light. They create a mental image of their surroundings using sound, a strategy called echolocation. Dolphins make high-frequency clicks by vibrating air in the nasal passages beneath the blowhole. The skull reflects the sounds as a beam. The melon in the dolphin's forehead focuses this beam. The dolphin listens for the echoes reflected back. The echoes are channeled through the lower jaw. The time delay between the sound being sent and its return, and the way it is distorted, tells the dolphin how far away an object is, whether it is moving, and what it is made of. The sound beam is like an X ray, able to penetrate through sand and living tissue. Air spaces, such as a fish's swim bladder, produce a particularly strong echo that the dolphin can home in on.

Dolphin intelligence

In captivity, animal trainers can teach dolphins to perform a wide range of tricks and to carry out complex tasks, such as retrieving specific objects when told to. In the wild, dolphin behavior is flexible, and individuals can learn quickly from each other: for example, to pick up and use a natural sponge as a "tool" to protect the snout when digging for food. Dolphins and most toothed whales have a large brain relative to their body size. A bottle-nosed dolphin has a brain that is, relatively, just a little smaller than a human's (and twice the size of a chimp's). Dolphin brains have a high degree of cerebral folding. This reflects the fact that their brain must process a fast stream of sound and visual information about their surroundings. Also, dolphins lead rich social lives and have to interpret each other's communications and behavior. Cerebral size and folding are good indicators of intelligence.

▶ *Captive dolphins can learn a range of tricks. These are leaping in unison.*

Circulatory and respiratory systems

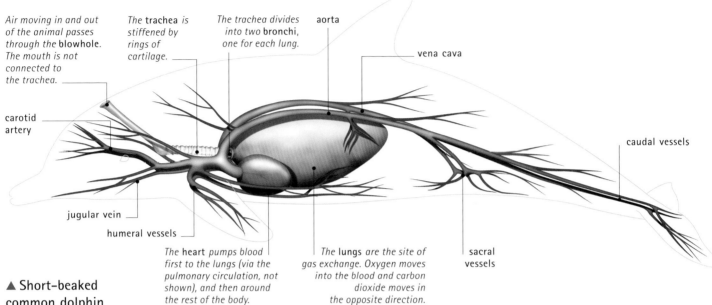

Air moving in and out of the animal passes through the **blowhole**. The mouth is not connected to the trachea.

The **trachea** is stiffened by rings of cartilage.

The trachea divides into two **bronchi**, one for each lung.

aorta

vena cava

carotid artery

caudal vessels

jugular vein

humeral vessels

The **heart** pumps blood first to the lungs (via the pulmonary circulation, not shown), and then around the rest of the body.

The **lungs** are the site of gas exchange. Oxygen moves into the blood and carbon dioxide moves in the opposite direction.

sacral vessels

▲ **Short–beaked common dolphin**
Important features of a short-beaked dolphin's circulatory system. Note the countercurrent systems that rewarm cooled blood returning from the fins, flippers, and flukes before it reaches the body core.

CONNECTIONS

COMPARE the dolphin's countercurrent heat exchanger with the countercurrent system in the gills of a **HAMMERHEAD SHARK** and a **TROUT**.

COMPARE diving adaptations of toothed whales with those of a **PENGUIN** and a **SEAL**.

Dolphins have a four-chamber heart that pumps blood through a double circulation system (the main and pulmonary circulations). Arteries with thick, muscular walls carry blood under high pressure away from the heart toward other organs. On the way the arteries divide into smaller and smaller blood vessels until, in the tiniest vessels, the capillaries, substances are exchanged with the surrounding tissues. The blood releases oxygen and food substances and takes up carbon dioxide and other wastes. Inside an organ, capillaries join together to form first venules (small veins) and then larger veins. Large, thin-walled veins carry the blood back to the heart under low pressure. In the pulmonary circulation, the lungs recharge the blood with oxygen, and waste carbon dioxide is removed.

Getting oxygen

Like all other mammals, whales breathe in air through nasal passages (a blowhole is a whale's nostril), in which the air is warmed before it travels down the trachea to reach the lungs. In the lungs, the inhaled air enters millions of tiny air sacs called alveoli. These have a large surface area; and there is a thin tissue barrier, just one

IN FOCUS

A heat exchanger

Whales maintain their body temperature at 97–99°F (36–37°C), but surface seawater can be very much cooler. At depths below 4,000 feet (1,200 m) it can plunge to 36–39°F (2–4°C). Even with an insulating layer of blubber under the skin, deep-diving whales are in danger of losing too much heat, particularly through the flippers, dorsal fin, and tail. A remarkable arrangement of blood vessels helps prevent this. Arteries that carry warm blood to the extremities surround the veins that carry cool blood back. This leads to an exchange of heat from arterial to venous blood. Such an arrangement is called a countercurrent heat exchanger. It helps ensure that the core temperature of the whale is not reduced significantly by incoming chilled blood returning from the extremities.

cell thick, between the inside of each alveolus and a system of capillaries that surrounds them. Oxygen diffuses from the alveoli into the bloodstream. There it binds to a protein

Avoiding the bends

In humans, there is a risk associated with scuba diving. At high pressures, nitrogen in inhaled air readily dissolves into the bloodstream. Because human divers continue to breathe underwater, nitrogen can accumulate in considerable quantities in the blood. When the diver rises and the pressure of the water reduces, there is a danger that the dissolved nitrogen will "fizz out," like the bubbles released by opening a bottle of cola. Nitrogen bubbles in the blood cause serious problems. In a condition called decompression sickness, or "the bends," bubbles of gas accumulate in the gaps between joints or in the brain, causing agonizing pain and sometimes death. Diving whales avoid this problem by ensuring that little nitrogen dissolves in the blood. They do not breathe underwater, and the gas in their lungs is not allowed to pass into the bloodstream. When whales dive with their lungs full of air, the increasing pressure forces their lungs to collapse. This drives air into the trachea and nasal passages, where gases cannot pass to or from the bloodstream.

pigment inside the red blood cells called hemoglobin. Hemoglobin carries the oxygen and releases it wherever it is needed. Meanwhile, waste carbon dioxide moves from the blood into the alveoli. The lungs expel the carbon dioxide when the whale surfaces and breathes out. Whales can empty and refill their lungs much more rapidly than humans; typically, in less than two seconds.

The cetacean circulatory and respiratory systems have many adaptations for diving. Even though all cetaceans breathe air, they can still remain underwater for extended periods. In the case of sperm whales and some beaked whales, this can be for more than two hours. Even small cetaceans, such as common and bottle-nosed dolphins, can remain submerged for up to eight minutes without breathing. When diving, whales reduce their demand for oxygen. Their heartbeat slows, and the amount of blood pumped around the body decreases. Smaller arteries that deliver blood to the skin, gut, and extremities constrict, reducing the blood flow; however, arteries supplying blood to vital organs such as the brain remain wide open. This allows oxygen to be delivered only to the organs that need it most. A pigment called myoglobin in the muscles stores oxygen; it releases its cargo during a dive.

Whale lungs are efficient oxygen extractors. Whales remove more than 80 percent of the oxygen in the air they inhale, as opposed to the 20 percent that human lungs can manage. When a whale surfaces after a dive, it explosively breathes out a cloud of air that is visible as a spout. The spout is a fine spray of water droplets that condenses from the whale's warm, humid breath.

The spermaceti

Sperm whales are deep divers. They have a structure inside their head called the spermaceti organ, which is filled with a waxy liquid. During diving, arteries contract to divert blood away from this organ. The spermaceti cools and becomes denser, helping the whale descend. Before surfacing, blood is sent back to the organ, so the spermaceti warms and becomes less dense, helping the whale ascend. The spermaceti organ also helps focus sound waves, like the melon of other toothed whales.

▼ *Dolphins do not breathe through their mouth. Instead, they breathe through a single crescent-shaped nostril, which has moved over millions of years to the top of the head. There it forms a blowhole. Muscles attached to the skull hold the blowhole closed when the animal is submerged; the blowhole opens when it visits the surface.*

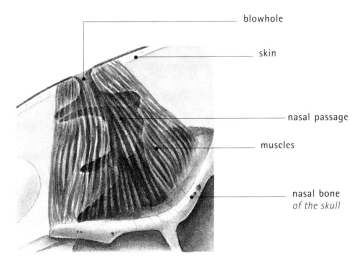

blowhole

skin

nasal passage

muscles

nasal bone
of the skull

Digestive and excretory systems

M̲ost toothed whales are fish eaters, although they are quite opportunistic and will take other available prey, such as squid and crustaceans. All toothed whales can use echolocation to home in on their prey. The arrangement of teeth reflects the whale's main food. Orcas have large, backward-curving, interlocking teeth that can snatch wriggling fish and bite chunks out of medium-sized whales.

Most dolphins have many small, pointed teeth to grasp slippery fish, which are swallowed whole. Risso's dolphin has a few teeth in the lower jaw only, which help it suck in and swallow squid. Sperm whales have a row of large, rounded teeth in the lower jaw that allow them to capture giant squid up to 50 feet (15 m) long. Porpoises have chisel-shaped teeth that work together like scissors to slice through their fish prey, which are large in proportion to their mouth.

Digestion

Toothed whales swallow their prey whole or in large chunks. After swallowing, the food travels down the esophagus to arrive at the stomach, a muscular sac where digestion (the process of breaking down food) starts in earnest. A whale's stomach has three compartments; it is more complex than that of other meat-eating mammals, such as humans, cats, and dogs. These types of animals have only one stomach chamber.

Muscular churning in the first part of the whale's stomach, the forestomach, grinds up the food. This process is called mechanical digestion. The resulting mush, called chyme, is squirted into the second chamber, the main stomach, where chemical digestion begins. The walls of the main stomach secrete hydrochloric acid to reduce the pH (increase the relative acidity) and protein-digesting enzymes to chemically break down the food. The walls of the third stomach compartment, the pyloric stomach, secrete fat-digesting enzymes, more protein-digesting enzymes, and an alkaline fluid that neutralizes the acidity of the main stomach juices.

Liver and pancreatic secretions

The partly digested food now enters a short tube called the duodenum, the first part of the small intestine. The pancreatic duct (leading from the pancreas) and the bile duct (from the liver) empty their contents into this organ. Pancreatic juices contain a mixture of digestive enzymes for different food types—proteins, fats, and carbohydrates. Bile is a liquid that contains bile salts. They emulsify (break up) globules of fat, making them smaller and easier to digest.

▼ **Short-beaked common dolphin**
The gut and excretory system (female). The kidney produces urine, which leaves the body at the urogenital opening. This opening also has other functions. In females, sperm is introduced there by a male, and young are born through it.

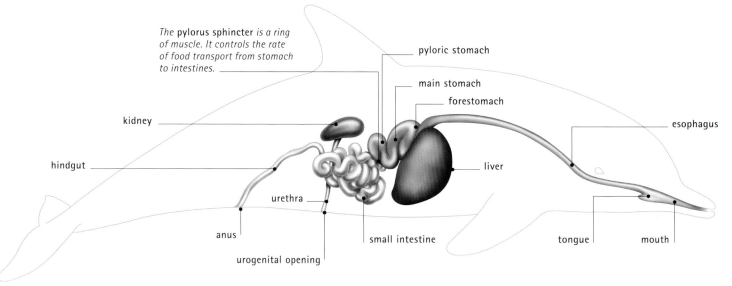

*The **pylorus sphincter** is a ring of muscle. It controls the rate of food transport from stomach to intestines.*

pyloric stomach

main stomach

forestomach

kidney

esophagus

hindgut

liver

urethra

anus

small intestine

tongue

mouth

urogenital opening

CLOSE-UP

Valuable vomit

Ambergris is a waxy substance that accumulates in the intestines of sperm whales. Tough bits of prey, such as squid beaks, collect in the whale's digestive tract. A thick, dark gray liquid is released by the intestine walls that coats this irritating material. The liquid hardens to form ambergris. Sometimes a whale vomits a lump of ambergris, which floats on the water surface.

Despite its strange origins, ambergris has a fragrant smell and was once highly prized by whalers. They sold it to perfume makers, who used it as a fixative in cosmetics.

PREDATOR AND PREY

Hunting as a team

Many toothed whales cooperate when hunting. Bottle-nosed dolphins often work together to drive a shoal of fish into shallow water. Boutu river dolphins may even drive the fish onto the shore before grabbing individual morsels and wriggling back into the water.

A pod of killer whales, or orcas, corrals a school of herring into a small area. The whales may emit high-pitched sounds to help them herd the terrified fish.

Members of the killer whale pod then take turns to drive through the tightly packed school of herring, snatching fish as they go.

Absorption and expulsion

Further chemical digestion takes place in the final section of the small intestine, the ileum, the walls of which release more digestive enzymes. Water and digested food are absorbed in the ileum and in the large intestine, which follows the small intestine. Indigestible waste is stored temporarily in the last part of the large intestine, the rectum, before being expelled through the anus.

Small and large intestines together form an immensely long tube, up to 500 feet (150 m) long in an adult sperm whale. The walls are highly folded and rich in blood vessels. This tube creates a very large surface area across which water and digested food substances can be absorbed efficiently, so little is wasted.

Expelling salts

Toothed whales, like other mammals, have a concentration of salts in their blood that is lower than the saltiness of seawater. Whales must take in water to keep their blood diluted, but at the same time they must get rid of excess salts; otherwise, their blood and tissues would become too salt-rich. Whales gain most of their water from the food they eat.

The kidneys remove excess salts from the bloodstream along with the by-products of cellular activity and other wastes. The kidneys filter the blood, reabsorb what is useful, and allow the rest to pass out of the body in urine. The bladder temporarily stores the urine until it is expelled to the outside through the urethra.

Whale kidneys are only moderately efficient at expelling salts. Thus whales have to excrete large quantities of water along with the salts. They have to replace the lost water through feeding rather than drinking.

Reproductive system

▶ **Short-beaked common dolphin**
Details of the male reproductive system.

Sperm is produced in the **testes,** *a pair of organs kept cool by blood routed from the dorsal fin. Sperm production increases during the breeding season.*

Tiny channels exit the rear of the testes. These merge to form the **epididymis,** *a coiled duct inside which sperm matures.*

The **vas deferens** *channels sperm from testes to the penis.*

pelvic bone

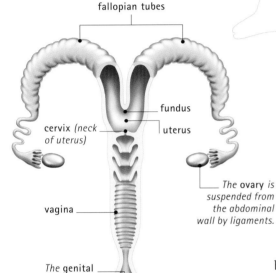

fallopian tubes

fundus

cervix (neck of uterus)

uterus

The **ovary** *is suspended from the abdominal wall by ligaments.*

vagina

The **genital slit,** *through which the calf is born.*

anus

The **genital slit,** *through which the penis protrudes during copulation and also some social interactions.*

The **penis** *is kept inside the body but it can quickly be everted.*

The **penis retractor muscle.** *Muscular control gives the penis a degree of flexibility.*

▲ **Short-beaked common dolphin**
The female reproductive system. Usually only one egg is released at a time.

Toothed whales have a similar reproductive system to that of placental land mammals, although adaptations have evolved that allow mating, birth of the calf, and suckling to take place underwater. To keep the body streamlined, male cetaceans have an internal penis and testes. The penis normally lies coiled inside the abdomen and is extended through a genital slit before mating. Muscles attached to the vestigial pelvis make the penis quite mobile, and it is used as a sensory organ, not just for mating but in other social encounters as well.

▶ *Before mating, male and female dolphins stroke each other with their flippers. Copulation takes place as the dolphins swim together belly to belly.*

In most toothed whale species, including dolphins, pairs practice considerable touching and rubbing before mating. Copulation, the process whereby the male's penis is inserted into the female's genital slit and sperm is released, is short-lived in dolphins. It typically lasts about one minute, but it may be repeated many times. Toothed whales are either promiscuous (males and females having many partners, as is the case in most dolphins) or polygynous (males mate with several females, as in sperm whales and narwhals).

Some adult male bottle-nosed dolphins form friendship alliances to gain access to mates and to keep other males at bay. The males within the alliance all mate with each female they successfully court; sperm from the different males compete inside the female to reach the ovum (egg cell) first. Male narwhals use their tusks to joust with each other for access to a particular female. Female toothed

GENETICS

Who are the parents?

Scientists can discover who the father of a dolphin calf is by taking tissue samples from the calf and matching its DNA with that of the local males. Researchers collect tissue samples using a small dart to remove a plug of skin. In some toothed whales, such as sperm whales, individuals slough off fragments of skin when they dive, and researchers collect these for DNA study.

IN FOCUS

Suckling young

A female dolphin's mammary glands are located on either side of the genital slit. They are normally hidden, but when the area is nuzzled by the calf, muscles contract to force a nipple to emerge from one of the mammary glands. Milk is then squirted directly into the calf's mouth. The calf does not need to suck, so it can drink underwater. Whale milk is much richer in fat than human or cow's milk. This helps the calf lay down a thick layer of insulating blubber within a matter of weeks.

whales often exercise their right to choose a mate by signaling to a male that they are receptive. Females signal behaviorally or by releasing chemical attractants in their urine.

Pregnancy and birth

Typically, a female dolphin's ovaries (egg-producing organs) release only one ovum at at a time. So if the female mates successfully only one calf is normally born. The gestation period (time from fertilization to birth) is 10–12 months in most toothed whales, but can be between 16 and 18 months in sperm whales.

A whale calf is born tail-first, so the young whale is supplied with oxygen through the placenta until the very last moment. Then the calf's blowhole emerges from the mother's genital slit, and the newborn is guided swiftly to the surface by its mother to take its first breath. This arrangement reduces the chances of the calf's drowning during birth. Calf and mother are linked by a short umbilical cord, connected to the placenta, but this severs naturally at a weak spot shortly after birth. The mother expels the placenta an hour or so later.

Several months before giving birth to her calf, a dolphin chooses another experienced female to act as a "midwife." The midwife looks after the expectant mother and helps her through the birth. The two adult dolphins move away from the rest of the school at this time. A strong bond often develops between the two female dolphins.

Growing up

Feeding a calf is energetically demanding. Also, a long time is needed to teach the calf the hunting and social skills it must have before it can fend for itself. For these reasons, a female usually produces a calf only every two to three years. Bottle-nosed dolphin calves begin to take solid food after about six months but are not fully weaned off milk until they are 18 to 20 months old.

The age of sexual maturity varies depending on species and gender. In dolphins, it is 5 to 14 years for females and 9 to 14 years for males. Female sperm whales typically mature at about 9 years old; males at 15 to 20.

TREVOR DAY

FURTHER READING AND RESEARCH

Mead, James G. and Joy P. Gold. 2002. *Whales and Dolphins in Question: The Smithsonian Answer Book.* Smithsonian Books: Washington DC.

▲ *Dolphins are unusual in that they enjoy play throughout their lives. Most animals play only when they are young. Play helps reinforce social bonds within the dolphin school.*

▼ *Attended by a helpful midwife dolphin, a female gives birth tail-first to her calf.*

Dragonfly

PHYLUM: Arthropoda CLASS: Insecta ORDER: Odonata
FAMILIES: Aeshnidae and several others

The 5,500 species of dragonflies and damselflies form the insect order Odonata. Throughout their lives, odonates (dragonflies and damselflies) are fierce predators. As nymphs (young) they catch insect larvae, worms, and even tadpoles. As adults, they catch small insects on the wing.

Anatomy and taxonomy

The Odonata are an ancient group; recent research suggests that they form a "sister" group to all other insects (except mayflies, which form a separate branch of the family tree).

● **Animals** All animals are multicellular and depend on other animals for food. They differ from other multicellular life-forms in their ability to move around (generally using muscles) and their rapid response to stimuli.

● **Arthropods** The Arthropoda have segmented bodies and a tough outer skin called an exoskeleton, which protects the internal organs and serves as an attachment point for muscles. Arthropods have pairs of jointed appendages that form structures such as legs and antennae. Internally, all arthropods have a ventral (running along the underside) nerve cord and a dorsal (running near the top surface) blood vessel.

● **Uniramians** Legs and other appendages of uniramian arthropods form a single structure—they are unbranched. Other arthropods, such as crustaceans, are biramous—their legs have an outer branch, which often forms a gill, and an inner branch used for walking. Most adult uniramians breathe through tracheae, tubes that take oxygen directly to the body cells.

● **Hexapods** Hexapods are six-legged arthropods. The vast majority of hexapods are insects, but there are three small noninsect hexapod groups. Noninsect hexapods, such as springtails and silverfish, do not have wings or antennae, and their mouthparts are kept in a pouch on the underside of the head.

● **Insects** Adult insect bodies are divided into three main sections: the head, thorax, and abdomen. The head bears mouthparts, antennae, and compound eyes that contain many individual lenses. The thorax contains three

▶ *This family tree shows the three groups of living odonates within the order Odonata.*

Animals
KINGDOM Animalia

Arthropods
PHYLUM Arthropoda

Uniramians
SUBPHYLUM Uniramia

Insects and allies
SUPERCLASS Hexapoda

Insects
CLASS Insecta

Dragonflies and damselflies
ORDER Odonata

Damselflies
SUBORDER Zygoptera

Dragonflies
SUBORDER Anisoptera

Anisozygopterans
SUBORDER Anisozygoptera

Biddies, Clubtails, and Graybacks

Darners
FAMILY Aeshnidae

Skimmers

Emperor dragonfly
GENUS AND SPECIES *Anax imperator*

208

segments: the prothorax, which bears a pair of legs; and the mesothorax and metathorax, each of which bears a pair of legs and, usually, a pair of wings, too. However, there is enormous diversity of insect anatomy. For example, some of the largest groups have only one pair of wings, and some others are wingless. The abdomen contains the bulk of the internal organs. It is formed by up to 11 segments.

Damselflies and dragonflies Adult odonates have a large head with biting mouthparts and compound eyes. The front of the head bears three light-sensitive structures called ocelli, and a pair of short, bristlelike antennae. The wings are large and do not beat together as one as in most other insects. The abdomen is long, with 10 visible segments. Odonate nymphs have an extendible lip on their head called a labial mask, which is used to grasp prey.

Damselflies Damselflies have a long, slim, cylindrical abdomen, and their wings are more or less equal in size. When they land, damselflies hold their wings above the body. The compound eyes are widely separated and do not meet. Females have a long ovipositor (egg tube) under the ninth and tenth abdominal segments. Damselfly nymphs are slender and have three feathery gills that jut from the rectum at the rear of the insect.

Anisozygopterans Although in the ancient past this suborder was numerous, just two species survive. They are intermediate in appearance between the other two odonate groups. Anisozygopterans have wings like those of a damselfly but a body like that of a dragonfly. Anisozygopteran nymphs are like dragonflies, with gills inside the rectum.

Dragonflies The head of a dragonfly is rounded, with massive compound eyes touching or even fused at the midpoint. Dragonfly hind wings are broader than the forewings and lack hinges at the base, so they are held flat

▲ *The emperor dragonfly. Unlike damselflies, dragonflies rest with their wings outstretched.*

when the insect is at rest. The nymphs are short and bulky, and do not have external gills; instead, their gills line the walls of the rectum. Oxygen-rich water is drawn into the rectum before being pumped out again. The nymphs use this to jet-propel themselves through the water.

Darners Most of the world's larger dragonflies are called darners. They are powerful fliers, and a few species, such as the green darner, even migrate long distances. The compound eyes of darners are fused. Darners often have colorful bands on their wings and body.

Emperor dragonfly Living throughout Europe, Africa, and parts of Asia, the emperor is a large dragonfly. The adult male has a sky-blue abdomen with a black stripe running along it; the thorax and head are bright green. The female's body is bright green. The nymphs are yellowish green.

FEATURED SYSTEMS

EXTERNAL ANATOMY Coated by a tough cuticle that must be molted to allow growth, dragonflies have three main sections: the head, the thorax, and the abdomen. They also have six legs, two pairs of wings, and a pair of large compound eyes. *See pages 210–213.*

INTERNAL ANATOMY The digestive system runs from mouth to anus. Other internal structures include the tracheal system, dorsal vessel pumps, and muscles. *See pages 214.*

NERVOUS SYSTEM Dragonflies have a tiny brain connected to a long nerve cord that stretches to the tip of the abdomen. Much brainpower is devoted to interpreting visual information. *See page 215.*

CIRCULATORY AND RESPIRATORY SYSTEMS
Hemolymph is shunted around the body by a long dorsal vessel with the assistance of several smaller pulsatile organs. Oxygen is not carried in the blood but is taken directly to cells by the tracheal system. *See page 216–217.*

REPRODUCTIVE SYSTEM A male dragonfly holds the female by the head to form a "wheel" of mating insects. The male may remove the sperm of rival males from the female's body. *See pages 218–219.*

External anatomy

Like other adult insects, a dragonfly has a body divided into three sections: the head, thorax, and abdomen. A light yet extremely tough outer coating called the exoskeleton covers the body. The exoskeleton has many important functions. It serves as an attachment for muscles, protects the insect from predators, helps the insect retain water, and contains pigments for coloration.

The exoskeleton is formed by a sheet of cells called the epidermis, which is supported by a membrane called the basal lamina. The cells secrete an outer skin called the cuticle. The cuticle has two main layers. The innermost of these (the procuticle) contains an important chemical, chitin.

The importance of chitin

Chitin is a sugar that forms long chains of molecules. These lie embedded in a matrix of proteins. The chains link to form structures called microfibrils that give the procuticle strength. The procuticle is further strengthened by the arrangement of the microfibrils: alternating layers of microfibrils orient at different angles.

The outer part of the procuticle is called exocuticle. Chitin in the exocuticle is sclerotized, or hardened; after each molt, the chitin reacts with chemicals called quinines to make it tougher. Chitin in the inner procuticle layer, the endocuticle, is not hardened in this way.

▶ **Emperor dragonfly**
Typical of dragonflies are the long, thin, brightly colored abdomen; very large compound eyes; and two pairs of independently movable wings.

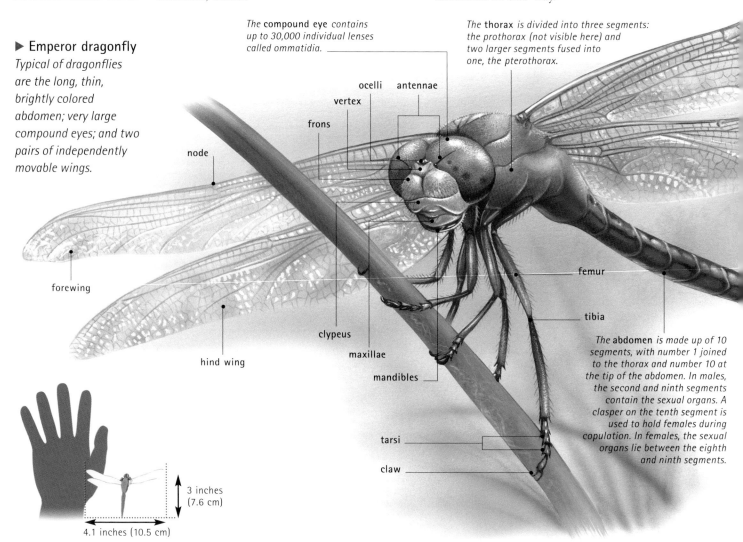

The **compound eye** contains up to 30,000 individual lenses called ommatidia.

The **thorax** is divided into three segments: the prothorax (not visible here) and two larger segments fused into one, the pterothorax.

ocelli   antennae
vertex
frons
node
forewing
hind wing
clypeus
maxillae
mandibles
femur
tibia

The **abdomen** is made up of 10 segments, with number 1 joined to the thorax and number 10 at the tip of the abdomen. In males, the second and ninth segments contain the sexual organs. A clasper on the tenth segment is used to hold females during copulation. In females, the sexual organs lie between the eighth and ninth segments.

tarsi
claw

3 inches (7.6 cm)
4.1 inches (10.5 cm)

▶ EXOSKELETON

All arthropods have an external skeleton called an exoskeleton. The exoskeleton provides support and protection for the dragonfly. However, the exoskeleton is inflexible and must be shed in a process called molting to allow the dragonfly to grow. This process is enabled by chemicals called hormones that are released from the molting glands.

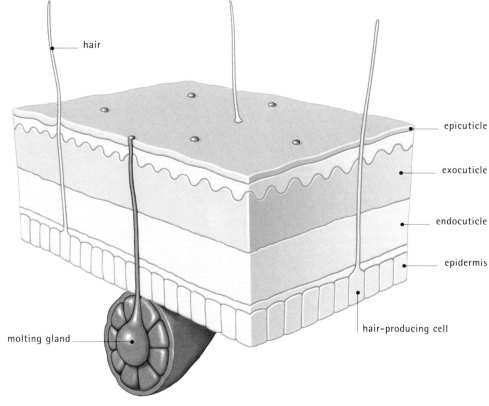

hair

epicuticle

exocuticle

endocuticle

epidermis

hair-producing cell

molting gland

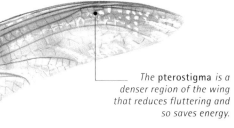

*The **pterostigma** is a denser region of the wing that reduces fluttering and so saves energy.*

claspers

A very thin layer called the epicuticle lines the procuticle. The epicuticle is coated with wax, which keeps the insect waterproof. Secretory cells called oenocytes, which lie between the epidermis and the basal lamina, produce the wax.

Wonderful wings

The toughness of the exoskeleton allowed insects to evolve a feature without parallel in the invertebrate world—wings. Wings are formed by two layers of exoskeleton, with hemolymph-filled veins sandwiched between. The veins form patterns that help biologists tell species apart. Odonate wings also have notches called nodes on their leading edges.

Dragonfly wings are light yet must be tough enough to withstand strong bending forces on the downstroke. During the upstroke, however, they need to be flexible and bend toward the body. Furrows on the wing allow the flexibility necessary for this to happen.

Near the front edge of each wing is a dark spot, the pterostigma. This relatively dense material influences the movement of the whole

---

COMPARATIVE ANATOMY

Comparing insect wings

Dragonflies are among the most aerobatic insect fliers. Unusually for insects, their wings beat independently—one pair moves up as the other goes down. Most insect wings are coupled, so the forewings and hind wings beat as one. The wings of many butterflies are coupled by a hook on one wing and a catch on the other; bees have a row of forewing hooks that link to a fold in the hind wing; bug wings are coupled by rows of bristles.

Some groups of insects have only one pair of wings. The house fly's hind wings have evolved (changed over very long periods of time) into structures called halteres that aid balance in flight. A group of insects called strepsipterans also have halteres, though theirs have evolved from forewings. The forewings of beetles form tough wing cases called elytra that protect the delicate hind wings from damage.

wing. The pterostigma reduces fluttering and allows the dragonfly to glide for longer at higher speeds, reducing the amount of flapping needed. This in turn reduces the amount of energy used by the dragonfly.

The head

Vision is important for dragonflies, since they detect prey, mates, and rivals visually. Their large head bears massive compound eyes, each containing up to 30,000 individual lenses. There are also three pairs of tiny light-sensitive structures called ocelli on the head, as well as a pair of small antennae. At the front of the head is a pair of biting mandibles lined by sharp, spiny teeth. The shape of the head depends on the animal's sex; female dragonflies are grasped by the head during mating, and the shape of their head allows the male easy attachment.

The thorax

The midbody or thorax is divided into three segments, but the first of these, the prothorax, is small. The other two are much larger and are fused into one, called the pterothorax. All three segments give rise to a pair of legs, but the wings are attached to the second and third only. The legs are segmented, with articulations (joints) between them to provide flexibility. Dragonflies catch other insects on the wing. Their legs are important for grabbing larger prey; smaller insects are simply snatched into

IN FOCUS

How wing veins work

The wing veins supply the wings with nutrients. They form a slightly raised profile on the wing surface. This aids flight in an unusual way. When a dragonfly flaps its wings, swirls of air called vortices roll off them, forming a temporary rotating structure in the air called a wake. The wing veins form much tinier vortices. They increase the lift produced by the wing by sucking it upward. Increased lift allows the dragonfly to glide for longer periods between flaps and use less energy getting from place to place.

▼ *The delicate wings of a dragonfly are perfectly suited to providing the maximum lift with the minimum of energy.*

the toothy mouth. The legs are important for perching and for manipulating prey, but they are rarely used for walking.

The abdomen and its appendages

The abdomen has 10 visible segments. It is long and flexible and bears the sexual structures. Males have three claspers that extend from the last abdominal segment. They use their claspers to attach to a female during mating. Farther toward the head is the male's gonopore, through which sperm leaves the body. Before mating, the male transfers sperm from this organ to a structure on the second segment called the aedeagus, from where it can be ejected.

Female odonates have a gonopore between the eighth and ninth abdominal segments through which sperm enters. Female damselflies have a long structure called an ovipositor formed by a pair of sharp tubes covered by plates. Eggs are laid through the ovipositor. Most female dragonflies (including emperor dragonflies) do not have an ovipositor.

Pores and holes

A dragonfly's exoskeleton has tiny holes called spiracles that allow air to move in and out of the tracheal (breathing) system. Dragonflies have eight spiracles: one on each of the second and third thoracic segments, and the rest on the abdomen. Muscles hold the thorax spiracles shut; the elasticity of the cuticle opens them briefly when the muscles relax. Two hard rods that pinch the tracheae beneath the abdominal spiracles keep the tracheae closed most of the time.

Nymph anatomy

Dragonfly nymphs are a little like the adults, although they have some important anatomical differences. Like all young insects, they lack sex organs and wings. The wings develop from a pair of pads that appear after the fourth molt—nymphs must molt up to 20 times before adulthood. The five-segmented abdomen is shorter and more rounded than the slender adult version, and it is tipped by five short appendages that include sensitive appendages called cerci. Nymphs also use an extendible labial mask to catch prey, rather than the legs and mouthparts as in adults.

How molting happens

Because the cuticle is virtually unstretchable, young insects must molt it from time to time if they are to grow. Molting is regulated by a series of hormones (chemical messengers) in the hemolymph. First, the exocuticle breaks away. Then a liquid called molting fluid is released that contains enzymes. A new epicuticle forms before the enzymes get to work. They break down the old endocuticle, while a new procuticle is deposited beneath. The insect is now ready to lose its old skin, of which just the epicuticle and exocuticle remain. The insect increases the pressure of its hemolymph to fracture the old cuticle. The insect then pulls itself out headfirst. The insect's entire cuticle is shed during molting, including internal sections such as the linings of parts of the gut and tracheal system. The nymph expands swiftly before the new cuticle begins hardening.

▼ *When a dragonfly emerges from its final molt, it pumps hemolymph to expand its wings before flying.*

Internal anatomy

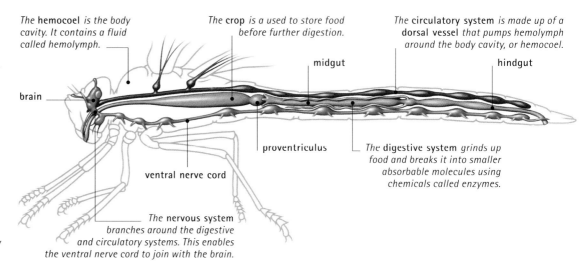

The **hemocoel** *is the body cavity. It contains a fluid called hemolymph.*

The **crop** *is a used to store food before further digestion.*

The **circulatory system** *is made up of a dorsal vessel that pumps hemolymph around the body cavity, or hemocoel.*

midgut

hindgut

brain

proventriculus

The **digestive system** *grinds up food and breaks it into smaller absorbable molecules using chemicals called enzymes.*

ventral nerve cord

The **nervous system** *branches around the digestive and circulatory systems. This enables the ventral nerve cord to join with the brain.*

▶ **Emperor dragonfly**
As with other insects, the internal structure of a dragonfly does not contain a system of veins for the transport of blood. Instead, a body fluid called hemolymph moves freely within the body cavity and supplies body cells with the products of digestion. The hemolymph is pumped around the body cavity by the dorsal vessel.

The digestive system is a prominent internal feature of dragonflies. From the mouth, food is pushed into the muscular pharynx, and from there through the esophagus into a storage structure called the crop. The crop allows the insect to control the rate at which food is digested. When nutrient levels in the hemolymph drop, the crop moves food into the midgut. Dragonflies do not chew their food with their teeth. Instead, food is ground up in the next part of the digestive tract, a muscular section called the proventriculus.

After grinding, food passes into the midgut. Unlike other parts of the digestive system, the midgut is not lined by cuticle. To prevent damage as food passes through, the cells that line the midgut release a membrane called a peritrophic envelope that wraps around the food. Midgut cells also release chemicals called enzymes. The enzymes are small enough to slip through pores in the peritrophic envelope, and they begin to digest (break down) the food into smaller molecules. Then, the products of digestion pass through the midgut cells and into the hemolymph.

Undigested material leaves the midgut and passes into the hindgut. At the junction of mid- and hindgut, excretory structures called Malpighian tubules empty urine. The mixture of undigested food and urine is excreted through the anus.

Other structures
There are many other structures inside a dragonfly's body. The dorsal vessel pumps blood around the hemocoel, or body cavity. Running through the body are tiny tubes that form the tracheal system. They deliver oxygen to cells and draw away carbon dioxide. Dragonflies have muscles that allow them to move. The muscles are attached to inward-jutting sections of the exoskeleton called apodemes.

COMPARATIVE ANATOMY

Beating wings

To power an upstroke, an insect pulls down with muscles on the tergite (top of the thorax). This makes the tergite flex downward, forcing the wings up. In most insects, a similar system powers the downstroke. Muscles squeeze the thorax, forcing the tergite upward and so bringing the wings down. Up to a third of the energy is saved on each wingbeat by the extraordinary elasticity of resilin, a chemical in the cuticle that helps it snap back into place. However, dragonflies power their downstrokes in a different way. They have powerful muscles that connect directly to the wings. These direct muscle attachments make flight less economical. They also prevent dragonflies from flapping their wings as quickly as many other insects can.

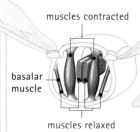

muscles contracted

basalar muscle

muscles relaxed

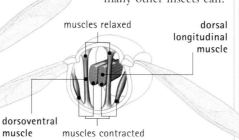

muscles relaxed

dorsal longitudinal muscle

dorsoventral muscle

muscles contracted

Nervous system

CONNECTIONS

COMPARE the position of the dragonfly nerve cord, which runs ventrally (along the underside), with that of a vertebrate such as a *HUMAN*, in which the nerve cord runs dorsally (along the back).

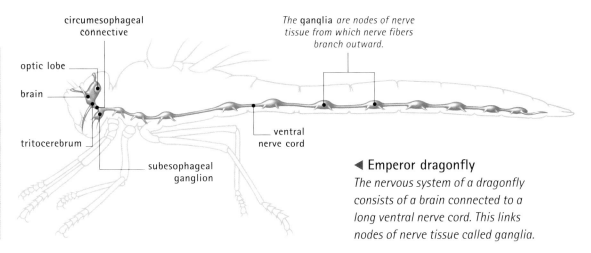

circumesophageal connective

optic lobe

brain

tritocerebrum

subesophageal ganglion

The ganglia *are nodes of nerve tissue from which nerve fibers branch outward.*

ventral nerve cord

◀ Emperor dragonfly
The nervous system of a dragonfly consists of a brain connected to a long ventral nerve cord. This links nodes of nerve tissue called ganglia.

An adult dragonfly depends on vision to identify food, mates, and predators. More than 80 percent of the dragonfly's brain, which lies just above the front of the digestive system, is devoted to interpreting such visual information. This takes place in the optic lobes. There are other lobes in the brain, such as the protocerebrum, which connects to the compound eyes and ocelli; and the deutocerebrum, which links to the antennae. The brain also contains two structures called the corpora allatum and corpora cardiaca. These structures release hormones into the hemolymph that control development and other body functions.

Connecting the body

At the base of the brain is another lobe, the tritocerebrum. This links to a pair of slender fibers called circumesophageal connectives, which skirt the esophagus before joining a bundle of nerves called the subesophageal ganglion. From this nodule, the pair of connectives extends into a long fiber called the ventral nerve cord, which runs along the underside of the body. In each body segment the nerve cord thickens to form a pair of structures called ganglia. They are linked by strips of tissue called commissures. Nerves radiate from the cord that control many of the body's functions.

IN FOCUS

How a dragonfly's eye works

A dragonfly's eye consists of up to 30,000 individual units called ommatidia. The cuticle that covers each ommatidium forms a lens, with a second lens, the crystalline cone, below. Pigments line the sides of each ommatidium so light cannot pass to another. A sensory cell extends from the cone to a nerve at the base of the ommatidium, which connects to the brain. The lenses focus light on the end of the sensitive cell. The signal it sends depends on the strength of the light hitting it. Thus dragonflies interpret the world as a series of dots of differing brightness.

compound eyes

Circulatory and respiratory systems

CONNECTIONS

COMPARE the open circulatory system of a dragonfly with the closed circulatory system of a mammal such as a *FRUIT BAT*.

COMPARE the tracheal system of a dragonfly with the lungs of a vertebrate such as a *HUMMINGBIRD*.

Like all other insects, dragonflies have an "open" circulatory system. The hemolymph (blood) is not contained within veins and arteries but instead moves around the body cavity, or hemocoel. Hemolymph is circulated by a long structure called the dorsal vessel. This runs beneath the top surface of the body, just below the exoskeleton, although it dips close to the gut at the front of the animal.

Dragonfly hearts

The dorsal vessel has two sections. The front part is a simple tube, while the rear part acts as a heart. The heart is perforated by slitlike openings called ostia; hemolymph moves out through some ostia (excurrent) and in through others (incurrent). A dragonfly's heart is divided into chambers separated by valves that lie behind the incurrent ostia. The valves stop hemolymph from flowing backward. Hemolymph moves forward as the heart contracts. It passes out from the dorsal vessel near the head, and also leaves through the excurrent ostia. The hemolymph moves down the body, aided by the movement of a structure called the dorsal diaphragm. When the heart relaxes, hemolymph moves back in through the incurrent ostia.

There are some other pumping vessels in addition to the heart called pulsatile organs. Because of this arrangement of dorsal vessel and pulsatile organs, a dragonfly effectively has

COMPARATIVE ANATOMY

Breathing underwater

Odonate nymphs need to breathe underwater. They do this with gills, a network of trachioles covered by very thin cuticle through which gases can diffuse. Damselfly nymphs have gills extending from the end of the abdomen. Dragonfly nymphs have internal gills. They lie in the rectum, forming a structure called the branchial chamber. Muscles force water from the chamber through the anus and out. Freshwater moves in as the branchial chamber expands through further muscle contractions and by the elasticity of the cuticle. Dragonfly nymphs can use this breathing system for jet-propelled movement. By forcing water backward from the anus, a nymph can move forward with a sudden burst of speed.

several "hearts." For example, the pulsatile organs in the second and third segments of the thorax help draw hemolymph from the wings.

Getting oxygen

Dragonfly hemolymph carries nutrients and wastes around the body. However, it does not carry oxygen to the cells. Instead, air enters the

▶ **Emperor dragonfly**
The function of a dragonfly's circulatory system is to pump nutrient-rich hemolymph around the body cavity. In contrast to the blood system of vertebrates, a dragonfly's hemolymph does not carry oxygen. Oxygen passes directly to cells through trachea. This system of narrow tubes leads to the external surface of the dragonfly's exoskeleton.

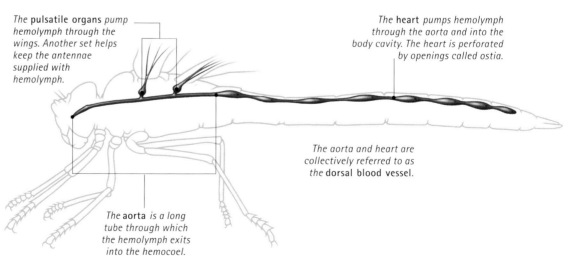

The **pulsatile organs** *pump hemolymph through the wings. Another set helps keep the antennae supplied with hemolymph.*

The **heart** *pumps hemolymph through the aorta and into the body cavity. The heart is perforated by openings called ostia.*

The *aorta and heart are collectively referred to as* the **dorsal blood vessel**.

The **aorta** *is a long tube through which the hemolymph exits into the hemocoel.*

EVOLUTION

Big beasts

The tracheal system is extremely efficient for animals the size of insects, but would be less efficient for larger animals. This places a limit on the maximum size of an insect, and explains why there are no elephant-size bugs, and why dragonflies will never evolve to be large enough to hunt humans. However, things were different in the ancient past. During the Permian period, around 250 million years ago there was much more oxygen in the atmosphere than today. Tracheae provided an effective means of obtaining oxygen for much larger organisms, so insects and other invertebrates could grow to much larger sizes. The largest known insect, *Meganeuropsis permiana*, lived around this time. It belonged to a group closely related to dragonflies called the Protodonata. *Meganeuropsis* had a wingspan of almost 30 inches (75 cm). *Meganeuropsis* shared the Permian forests with many other giant invertebrates, including a spider with a legspan of more than 20 inches (50 cm).

▲ Meganeuropsis permiana *is the largest insect known to have existed. A close relative of living odonates, it had a wingspan of almost 30 inches (75 cm).*

body through pores called spiracles that occur along the length of the thorax and abdomen. Air passes along a series of increasingly narrow tubes called tracheae, with the finest branches (the trachioles) connected directly to cells.

Oxygen moves into the cell at the tips of the trachioles, with carbon dioxide traveling in the opposite direction. The tracheae are lined with cuticle and cannot easily be compressed, so air sacs are dotted through the system. The pressure of the hemolymph is raised to squeeze the air sacs. This flushes out the system, allowing fresh air to move in.

▶ *A dragonfly nymph breathes using gills in its anus. Water taken into the anus for breathing can be quickly ejected to jet-propel the nymph forward.*

Reproductive system

To mate, a male dragonfly uses its claspers to grasp the head of a female. The female then curls her abdomen upward, so the couple forms a "wheel" shape. Like all other male insects, dragonflies have an exit point for sperm, or gonopore, on the ninth abdominal segment. Sperm is made in the testes and passes out through a duct called the vas deferens. However, male dragonflies have a set of secondary sex organs on the second abdominal segment. They include the aedeagus, a four-segmented structure that is used to introduce sperm into the female's body. Before searching for a female, the male "charges" the aedeagus with sperm by folding his abdomen. The sperm is then stored inside the aedeagus.

Copulation

The entry to the female's reproductive system lies between the eighth and ninth abdominal segments. The male aedeagus enters a structure inside the female called the bursa, and sperm is released. It is kept in the bursa and a pair of separate storage structures, the saclike spermatheca. The storage structures keep the male's sperm fresh until fertilization (fusion of egg and sperm) occurs.

The eggs are produced by the female's ovaries. The eggs pass along a duct that widens into a structure called the vagina, where fertilization takes place. A pair of sensitive cuticular plates lines the vagina. These plates are involved in the release of stored sperm from the spermatheca, which is then used to fertilize the eggs.

Laying eggs

Some female dragonflies have a pointed ovipositor, an extension of the plates that surround the entry to the bursa. Such dragonflies crawl beneath the surface of the water—often with the males still clinging to their heads—to lay up to several hundred fertilized eggs inside underwater plants or rotting wood. Species that lack ovipositors lay their eggs on or just below the water surface.

The nymphs generally have 12 instars (stages), each separated by a molt; it can be several years before development is completed. Then the nymphs climb out of the water and

▼ **Emperor dragonfly**
Male (left) and female (right) reproductive organs. The male dragonfly's aedeagus functions as a penis. This structure has widely different sizes and shapes, depending on the species. The number of sperm-removing appendages also varies.

The emperor dragonfly **aedeagus** *has four appendages (three visible here) that remove the sperm of rival males from the bursa of the female.*

appendages

Aedeagus

testes

sperm tubes

accessory glands

vas deferens

seminal vesicle

ejaculatory duct

Male

ovaries

spermatheca

oviduct

spermathecal gland

accessory gland

bursa

Female

PREDATOR AND PREY

Changing mouthparts

The mouthparts of an odonate change substantially during metamorphosis, the change from nymph to adult insect. In odonate nymphs the mouthparts form an unusual extendible anatomical feature called a labial mask. This is formed by the elongated postmentum and prementum (sections of the head), with the palps at the tip forming sharp pincers. At rest (below), the mask lies beneath the head. When prey comes close, the mask shoots out with astonishing swiftness. Just before the strike, a ring of muscle closes the anus, while other muscles squeeze the hemolymph, increasing its pressure. Meanwhile, muscles in the prementum pull the mask close to the head. The cuticle linking the two mask sections is elastic. When the muscles are released, the prementum shoots forward like a catapult.

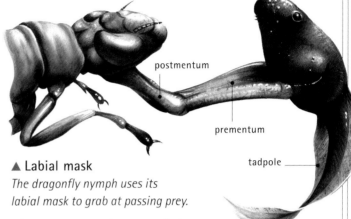

▲ **Labial mask**
The dragonfly nymph uses its labial mask to grab at passing prey.

The postmentum then extends due to the pressure of the hemolymph. When a nymph molts for the last time, its labial mask is replaced with adult feeding appendages, the maxilla and mandibles. These are lined with spiny teeth for subduing flying insect prey.

shed their skin for the last time. They wait for their wings to harden before taking flight to search for food and mates.

JAMES MARTIN

▼ *A male damselfly holds the head of a female with his claspers during mating and transfers sperm to her bursa.*

FURTHER READING AND RESEARCH

Blum, Mark. 1998. *Bugs in 3-D.* Chronicle Books: San Francisco, CA.

Jackson, T. and J. Martin (eds). *Insects and Spiders of the World.* 2003. Marshall Cavendish: Tarrytown, NY.

Sperm wars

Female dragonflies store sperm before fertilization takes place. This provides an opportunity for males to remove the stored sperm of rival males before fertilization. The final segment of a male dragonfly's aedeagus bears a variety of inflatable lobes and spikes. They may push older sperm away from the points of fertilization, or scoop out the sperm completely. Also, the movement of eggs inside the female during oviposition (egg laying) triggers sensory cells that lie on the cuticular plates. They cause the spermatheca to flush out any sperm they contain. In some damselflies, the entry of the aedeagus into the female stimulates the same cells, so any rival sperm is ejected. Such conflict between the sex cells of male animals is called sperm competition. Sperm competition is particularly fierce in odonates. That is why males guard the females so zealously after mating. Many species remain attached in the mating wheel for several hours.

Eagle

ORDER: Falconiformes FAMILY: Accipitridae
GENERA: *Aquila* and 21 or 22 others

The more than 60 species of eagles are predatory birds. Eagles have adapted for life in almost every terrestrial environment, from rain forests to polar deserts, and they live in every continent except Antarctica.

Anatomy and taxonomy

Scientists group all organisms into taxonomic groups based largely on anatomical features. Eagles belong to the family Accipitridae, together with hawks, Old World vultures, and kites. Along with falcons, caracaras, the osprey, and the secretary bird, eagles and hawks are members of the avian (bird) order Falconiformes.

● **Animals** All animals are multicellular and depend on other organisms for food. They differ from other multicellular life-forms in their ability to move around (generally using muscles) and respond rapidly to stimuli.

● **Chordates** At some time in its life cycle a chordate has a stiff, dorsal (back) supporting rod called the notochord.

● **Vertebrates** The notochord of vertebrates develops into a backbone made up of units called vertebrae. Vertebrates have a muscular system consisting primarily of muscles that are bilaterally paired (they lie on each side of an imaginary line of symmetry running along the body from head to tail). Vertebrate classes include the mammals, birds, reptiles, amphibians, and fish.

● **Birds** The second largest vertebrate group, birds evolved from reptilian ancestors more than 150 million years ago. Their most obvious identifying feature is the feathers that cover the body. Birds are bipedal (walk on two legs); they do not have teeth; and at least some of their skeleton is pneumatized, or hollow. Most birds are able to fly, and all are descended from flying ancestors.

● **Birds of prey** The order Falconiformes, or birds of prey, is characterized by a strong, sharply hooked bill; strong feet with sharp talons; long, strong legs; nostrils that open in a leatherlike cere (patch of skin at the base of the bill); and a predatory lifestyle. Birds of prey hunt a wide variety of prey or scavenge carrion (dead meat).

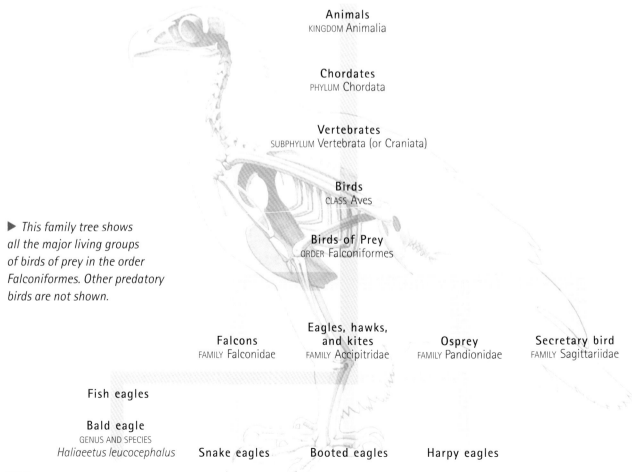

▶ *This family tree shows all the major living groups of birds of prey in the order Falconiformes. Other predatory birds are not shown.*

Animals
KINGDOM Animalia

Chordates
PHYLUM Chordata

Vertebrates
SUBPHYLUM Vertebrata (or Craniata)

Birds
CLASS Aves

Birds of Prey
ORDER Falconiformes

Falcons
FAMILY Falconidae

Eagles, hawks, and kites
FAMILY Accipitridae

Osprey
FAMILY Pandionidae

Secretary bird
FAMILY Sagittariidae

Fish eagles

Bald eagle
GENUS AND SPECIES
Haliaeetus leucocephalus

Snake eagles

Booted eagles

Harpy eagles

● **Falcons** Representatives of the family Falconidae occur in every continent apart from Antarctica. While sharing many anatomical features with other birds of prey, most falcons have relatively long and narrow wings and are well equipped for rapid flight in pursuit of prey.

● **Eagles** There are more than 60 species of eagles in 21 or 22 genera (groups of closely related species). Closely related to hawks, eagles have long, broad wings and a broad tail, allowing them both to soar and flap actively. Some eagles are very large, weighing up to 20 pounds (9 kg). Young eagles do not attain adult plumage until they are at least a year old; in larger species this may take five years.

● **Harpy eagles** This group includes some of the largest and most powerful eagles. There are six species that live in the forests of South America, Africa, and East Asia.

● **Booted, or true, eagles** Distinguished from other eagles by legs that are feathered down to their feet, this is a very varied group, including both the tropical hawk-eagles and the wide-ranging *Aquila* species. Several booted eagles have long crest feathers. There are 36 species in 9 genera.

● **Snake eagles** Generally smaller than other eagles, snake eagles live mostly in the savanna and forests of tropical and subtropical Africa, Asia, and Australia. There are 14 or 15 species in 5 genera.

● **Fish eagles** These birds live around the shores of lakes, rivers, and oceans, from the Arctic Ocean to the tropics, but not in South America. Fish eagles eat fish, taken alive or dead. There are 10 or 11 species in 2 or 3 genera.

● **Bald eagle** *Haliaeetus leucocephalus* is a large piscivorous (fish-eating) eagle that lives only in North America. Females are larger than males. The average wingspan of a

▲ *An adult bald eagle. Bald eagles feed mostly on fish, although small mammals, wildfowl, and carrion may also be eaten.*

bald eagle is 6 to 7 feet (200 to 235 cm), and the average weight of a female is 12 pounds (5.5 kg). An adult bald eagle is covered with brown feathers apart from its head, tail, and wing tips, which are white.

FEATURED SYSTEMS

EXTERNAL ANATOMY Eagles are generally large flying birds with a hooked bill and sharp talons. *See pages 222–224.*

SKELETAL SYSTEM An eagle's skeleton is adapted for flight, with light, hollow bones. It also has long, strong bones in its legs that are able to grasp and carry relatively heavy prey. *See pages 225–226.*

MUSCULAR SYSTEM Powerful muscles attached to the sternum (breastbone) control the movement of the wings, allowing eagles to fly powerfully. *See pages 227–228.*

NERVOUS SYSTEM Eagles have excellent vision that enables them to detect prey in the distance. *See page 229.*

DIGESTIVE AND EXCRETORY SYSTEMS Eagles catch prey using their talons and tear it up into small pieces with the beak. They do not digest all the food they swallow, but regurgitate fur, bones, and fish scales. *See page 230.*

CIRCULATORY AND RESPIRATORY SYSTEMS Eagles take in oxygen through their lungs, but their respiratory system also includes a series of air sacs, as well as hollow parts of some of the longer bones. *See page 231.*

REPRODUCTIVE SYSTEM Eagles reproduce sexually. Adult females generally lay a clutch of two eggs every year. The young reach sexual maturity after four years. *See pages 232–233.*

External anatomy

CONNECTIONS

COMPARE the bill of an eagle with that of a **PENGUIN** and a **HUMMINGBIRD.**

COMPARE the feet of an eagle with those of an **OSTRICH** and an **ALBATROSS.**

The feature of birds that distinguishes them from all other animals is the feathers that cover their body and head, and sometimes their legs as well. There are some other features that birds share with only a few other vertebrates. For example, they are bipedal: they walk on their hind limbs. Their forelimbs have evolved into wings that are used for powered flight. Although not all birds are able to fly, all have feathers, and all have wings. However, the wings of some flightless birds, such as ostriches, emus, kiwis, rheas, and the three species of cassowaries, are not well developed.

Size and shape

Eagles are medium-size or large birds. The biggest, the harpy eagle, weighs up to 20 pounds (9 kg). Eagles are accomplished fliers, and some can fly very long distances on migration. Others spend most of the hours of daylight soaring (that is, gliding with few wing beats) in search of prey. All eagles have two relatively long, and often broad, wings. Having wings with a large surface area enables eagles to spend long periods soaring high above the ground. This mode of flight uses only a fraction of the energy of flapping flight.

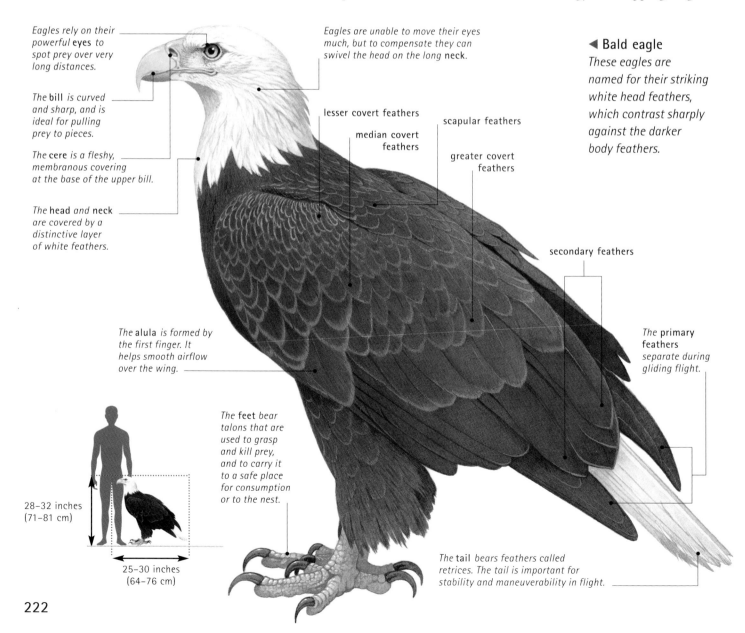

Eagles rely on their powerful **eyes** to spot prey over very long distances.

The **bill** is curved and sharp, and is ideal for pulling prey to pieces.

The **cere** is a fleshy, membranous covering at the base of the upper bill.

The **head** and **neck** are covered by a distinctive layer of white feathers.

Eagles are unable to move their eyes much, but to compensate they can swivel the head on the long **neck.**

lesser covert feathers

median covert feathers

scapular feathers

greater covert feathers

◀ **Bald eagle**
These eagles are named for their striking white head feathers, which contrast sharply against the darker body feathers.

secondary feathers

The **primary feathers** *separate during gliding flight.*

The **alula** *is formed by the first finger. It helps smooth airflow over the wing.*

The **feet** *bear talons that are used to grasp and kill prey, and to carry it to a safe place for consumption or to the nest.*

28–32 inches (71–81 cm)

25–30 inches (64–76 cm)

The **tail** *bears feathers called retrices. The tail is important for stability and maneuverability in flight.*

Why eagle wings have fingers

Birds and some other animals are able to fly because their wings act as airfoils. An airfoil's shape forces air to move more quickly across the top surface than the bottom surface. The resulting difference in pressure creates a force called lift, which pulls upward. Lift helps keep flying animals airborne. Individual feathers, too, have aerodynamic (lift-producing) properties. Shaped like individual airfoils, the flight feathers (the primaries and secondaries) are asymmetrical; vanes on the front, or leading, edge are shorter, thicker, and less flexible than those on the trailing edge. This stops the feather from twisting during the downstroke, which would cause a decrease in lift.

In birds like eagles the primaries extend from the wing tip like a set of fingers. During flight these point upward a little. Swirls of air moving around the separated primaries causes a decrease in another force, drag—this is the resistance of the air to movement through it. Because drag is reduced, less energy is used during flight. Separated primaries are of particular importance to birds that spend much of the time soaring, like vultures, eagles, and buzzards. Such birds often use thermals (rising air columns) to move upward. Once they leave the thermal, they glide slowly downward. The lower the drag, the longer the bird can remain airborne without having to flap or find another thermal. This gives them more time to search for food and saves energy.

The primary and secondary flight feathers of an eagle's wings are partially covered by layers of covert feathers. These coverts protect the flight feathers from damage and also help the birds remain airborne at low speeds. The largest eagles have a wingspan (the distance from the tip of one wing to the tip of the other when the wings are outstretched) of up to 10 feet (3 m). Even the small Congo serpent eagle has a wingspan of 33 inches (85 cm) despite having a body and tail length of just 20 inches (50 cm).

▶ Eagles are expert soarers. Their separated flight feathers help them glide for long periods without flapping.

Head and neck

An eagle's mouthparts are adapted for tearing and cutting flesh. Rather than toothed jaws, eagles have a hooked bill, the base of which is saddled by an unfeathered area called the cere. The two nostrils open in the cere, which often becomes brightly colored when the bird is ready to breed. The combined upper jaw and bill are called the maxilla. The lower jaw and bill are the mandible. The upper bill has sharp cutting edges, so when the lower

Eagle ancestors

The oldest falconiform fossils date from the Eocene epoch, around 50 million years ago, although the group probably appeared before the extinction of the dinosaurs 65 million years ago. Some eagles today are very large, but all are dwarfed by a species that died out less than 1,000 years ago. Haast's eagle lived on the islands of New Zealand. It had very robust legs, talons the size of a tiger's claws, and a skull 6.5 inches (16.4 cm) long. Haast's eagle preyed on giant flightless birds called moas—some fossil moa bones bear scars scratched by the claws of this giant eagle. Moas became extinct following climatic changes and intensive hunting by Maori peoples, who arrived on the islands from around the 10th century onward. Consequently, the moa's demise led to the disappearance of the giant eagle.

bill pushes against it the bill can slice through the flesh and even bone of its prey.

Eagles have two large eyes, each of which has three protective eyelids: the upper and lower eyelids close to meet at the middle of the eye. The third, a tough, opaque eyelid called the nictitating membrane, shuts when the eagle makes impact with moving prey. The nictitating membrane protects the eye while allowing some light in, and it also helps clean and moisten the surface of the eye.

Eagles have a relatively long neck, though this is not always an obvious feature, owing to its feathered covering. The long neck makes it easier for the bird to rotate its head while scanning for prey. This is particularly important for eagles, since their eyes are too large to rotate much in their sockets.

Legs and tail

Eagles use their legs and feet for terrestrial movement and for grasping prey. All eagles have four toes on each leg, three of which face forward; each toe has a sharp talon. Bald eagles and other fish eagles have spines called spicules on the underside of their toes that help them grasp slippery fish. While most species locate their prey from the air, others sometimes search for a meal when they are on the ground. For these terrestrial hunters, notably the snake eagles, their legs and feet are doubly important—as a means of locomotion and as weapons against their prey.

Eagles can be divided into two groups according to the feathering, or lack of it, on their legs. Bare-legged eagles include the fish eagles, snake eagles, and serpent eagles. "Booted" eagles include all those in the large genus *Aquila* and the hawk-eagles.

All eagles have a tail of 12 or 14 feathers, called the rectrices. These feathers vary in length according to the species. Several species of hawk-eagles have a particularly long tail that helps them maneuver at speed when flying through forest. In contrast, the bataleur—an eagle of the African savanna—has a tail so short that its feet extend beyond the tip when the bird is in flight.

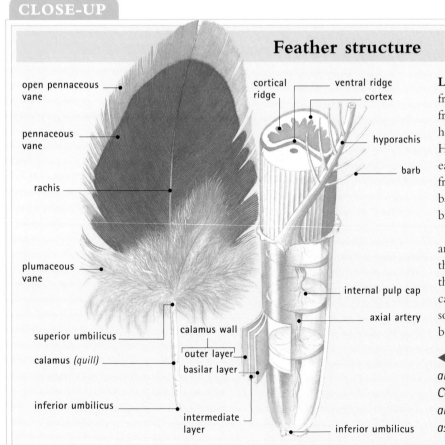

CLOSE-UP

Feather structure

open pennaceous vane

pennaceous vane

rachis

plumaceous vane

superior umbilicus

calamus *(quill)*

inferior umbilicus

calamus wall

outer layer

basilar layer

intermediate layer

cortical ridge

ventral ridge

cortex

hyporachis

barb

internal pulp cap

axial artery

inferior umbilicus

Like reptile scales, feathers are made from a protein called keratin. They grow from feather follicles in the skin. Each feather has a long central shaft called a rachis. Hundreds of parallel vanes branch off on each side. A vane is formed by a long barb, from which tiny barbules protrude. The barbules of adjacent vanes interlock, binding the barbs to form a strong sheet.

Feathers suffer a huge amount of wear and tear, so they must be replaced regularly through molting. While most birds molt all their feathers every year, some large eagles cannot complete their molt that quickly, so a second and even a third molt starts before the first has finished.

◀ *Contour feather structure with the anatomy of the quill, or calamus, in detail. Contour feathers include the flight feathers and those important for insulation, such as this one.*

Skeletal system

COMPARE the leg bones of an eagle with those of a **_HUMMINGBIRD_**.

COMPARE the wing bones of an eagle with the forelimb skeleton of a **_JACKSON'S CHAMELEON_**.

CONNECTIONS

The skull is composed of thin sheets of bone with tiny supporting struts. It is extremely strong to support the jaw musculature and protect the brain.

cervical vertebrae

The clavicle is one of a pair of bones that form the furcula, or wishbone.

coracoid

The sternum serves as an attachment for the largest muscles in the body, the pectorals.

humerus
carpometacarpus
radius
ulna

The wing bones that support the flight feathers.

thoracic vertebrae

vertebral rib
sternal rib

caudal vertebrae
pygostyle

carina (or keel)

femur

ilium ischium pubis

fibula

tibiotarsus

tarsometatarsus

The pelvis provides a sturdy support for the leg muscles and protects the reproductive organs.

basal phalanx
terminal phalanx
digit

▲ Bald eagle
Not only do an eagle's bones provide support and attachment points for muscles; they also form an integral part of the respiratory system.

For birds, a skeleton with both strength and lightness is key. An eagle's skeleton is strong and rigid to support all the stresses of the bird's active hunting lifestyle, while remaining light enough to allow active flight. Birds lack some of the bones possessed by other vertebrates, and other groups of bones are fused. The fusion of certain bones helps make the skeleton stronger. For example, fused bones in the central section of the backbone help the spine resist the bending and twisting forces created by flapping wings. Other fused bones that are separate in mammals include those near the wing tips. The bones of an eagle's pelvis are fused into a strong platform to which the leg, tail, and abdominal muscles are attached.

Eagle skeletons also lack some features that add weight in other vertebrates, such as a long, bony tail. An eagle's tail is relatively long, but only the base is bone; most of its length is feathers. As with other birds, most of an eagle's larger bones are hollow, and therefore lighter than solid bones. Such bones are said to be pneumatized, and they are linked to the respiratory system. Pneumatized bones are strengthened by supporting struts and are very strong for their weight.

Bone by bone

The main elements of an eagle's skeleton are the skull; the spinal column; the wing bones; the rib cage; the sternum, or breastbone; and the leg bones. All eagles have a skull with

▼ SKULL
Bald eagle

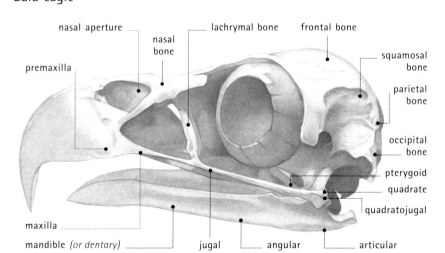

nasal aperture
nasal bone
lachrymal bone
frontal bone
premaxilla
squamosal bone
parietal bone
occipital bone
pterygoid
quadrate
quadratojugal
maxilla
mandible (or dentary)
jugal
angular
articular

Bones of the legs and feet

Perhaps the most striking features of an eagle's skeleton are the leg bones. The tibiotarsus and fibula are both long, while the femur, tibiotarsus, and tarsometatarsus are also thick. The bones of the feet—the metatarsals and digits—are strong, and the latter are hooked into sharp talons.

Almost all the thickness of an eagle's legs and feet is bone. All eagles have strong legs, and those of the harpy eagle are probably the strongest. Its legs are 4 to 5 inches (10–13 cm) in circumference; its foot span reaches 9 inches (23 cm), and its talons are 1.5 inches (3.8 cm) long. With these it is able to grab and kill adult monkeys from the canopy of the South American rain forest.

By contrast, the foot span of an African pygmy falcon, another member of the order Falconiformes, is just 1.5 inches (3.9 cm) across. The pygmy falcon does not usually catch prey larger than large insects or small lizards.

▲ African pygmy falcons have small feet with sharp talons. They use the talons to snatch small prey such as lizards from the ground. They also catch insects and, occasionally, small birds on the wing.

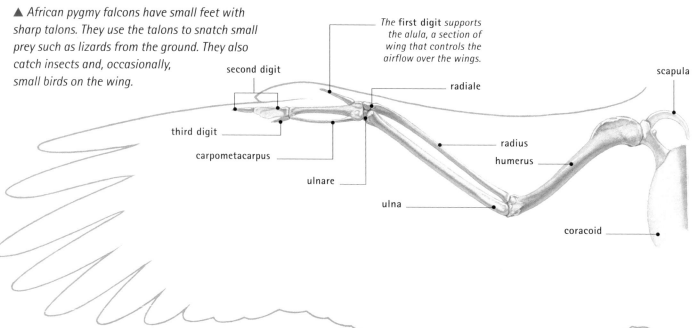

The first digit supports the alula, a section of wing that controls the airflow over the wings.

second digit

third digit

carpometacarpus

ulnare

radiale

radius

humerus

ulna

scapula

coracoid

▲ WING BONES

The wing bones must be strong to withstand the bending forces produced by flight, while remaining light. The bones are hollow to minimize weight. Strength is provided by a series of struts within the bones.

a large bill, used for tearing apart prey. The skull has a particularly large, hook-tipped upper jawbone (or maxilla); spacious orbits in which the very large eyeballs sit; and a supraorbital ridge of bone over each eye.

The vertebrae (spinal bones) of an eagle's spinal column can be divided into four groups: the cervical, thoracic, abdominal, and caudal vertebrae. Eagles have only 14 vertebrae in their neck, although most other birds of prey have 15. The pectoral girdle, which includes the sternum, coracoid, scapula, and clavicle, anchors the wings to the body. The sternum (breastbone) is large to support the powerful muscles that flap the wings.

The bones that support the eagle's wing feathers—the humerus, ulna, and radius—are longer than in most other birds. The "hand" of the wing skeleton comprises the carpometacarpus and the digits. They extend from the base of the radiale and ulnare bones, which attach in turn to the radius and ulna.

Muscular system

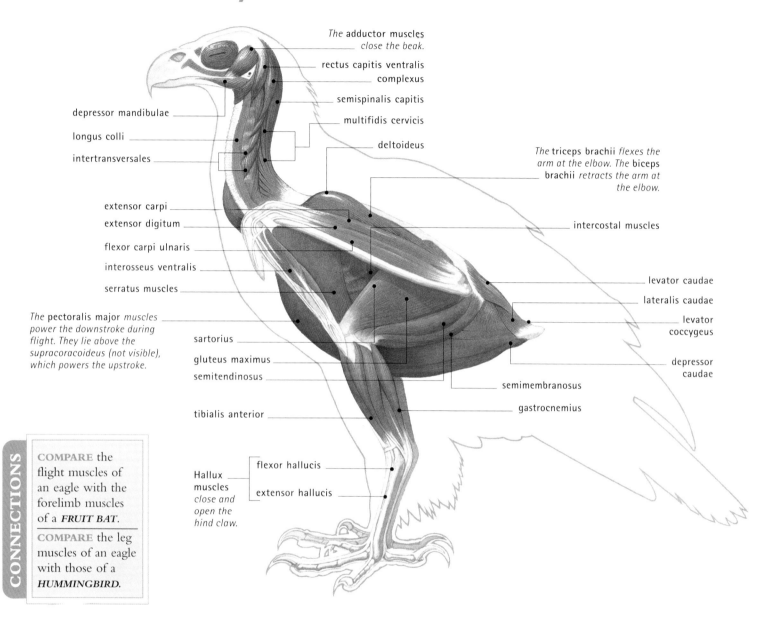

The **adductor muscles** *close the beak.*

rectus capitis ventralis

complexus

semispinalis capitis

multifidis cervicis

deltoideus

depressor mandibulae

longus colli

intertransversales

extensor carpi

extensor digitum

flexor carpi ulnaris

interosseus ventralis

serratus muscles

The **pectoralis major** *muscles power the downstroke during flight. They lie above the supracoracoideus (not visible), which powers the upstroke.*

sartorius

gluteus maximus

semitendinosus

tibialis anterior

Hallux muscles *close and open the hind claw.*

flexor hallucis

extensor hallucis

The **triceps brachii** *flexes the arm at the elbow. The* **biceps brachii** *retracts the arm at the elbow.*

intercostal muscles

levator caudae

lateralis caudae

levator coccygeus

depressor caudae

semimembranosus

gastrocnemius

CONNECTIONS

COMPARE the flight muscles of an eagle with the forelimb muscles of a *FRUIT BAT*.

COMPARE the leg muscles of an eagle with those of a *HUMMINGBIRD*.

An eagle's wing muscles are thin, tough bands of tissues. Their arrangement and size give many clues as to the bird's lifestyle. A large part of an eagle's mass is made up of the pectoralis major muscles. They power the downstroke of the wings; the upstroke is powered by a much smaller muscle, the supracoracoideus. The flight muscles attach to the humerus bone of each wing and the keel, or carina, of the sternum. The powerful and relatively heavy flight muscles are very close to the bird's center of gravity, allowing the greatest possible stability in flight.

If such a large proportion of the bird's weight were concentrated away from its center of gravity, it would be unable to fly.

The wing muscles

Muscles attached to different sections of the wing skeleton perform different functions. The muscles attached to the humerus control the position of the forearm. Some enable the wings to be extended (extensor muscles), and some close them (flexor muscles). Eagles lack the expansor secondarium muscles that most birds have in their wings. These muscles

▲ **Bald eagle**
The eagle's muscles are very powerful. Large muscles are required for flight; they must provide enough forward force to keep the bird airborne, even when carrying heavy prey. Strong leg and claw muscles are used for grasping prey.

227

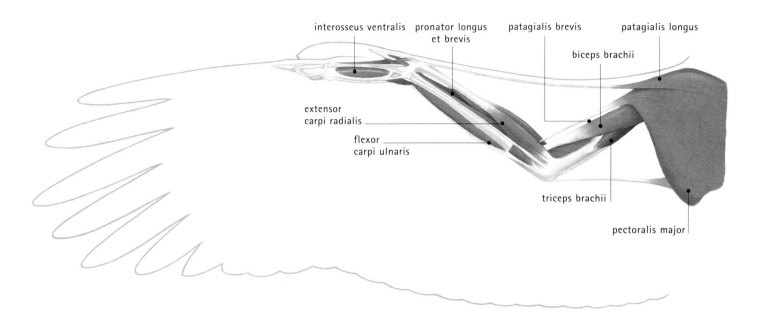

interosseus ventralis | pronator longus et brevis | patagialis brevis | patagialis longus | biceps brachii | extensor carpi radialis | flexor carpi ulnaris | triceps brachii | pectoralis major

▲ WING MUSCLES

Although most of the power for flight comes from the chest muscles, the wing muscles are important for retracting and extending parts of the wing through the wing beat. Without this movement, powered flight could not occur. Wing muscles also allow the angle and area of the wings to be changed, altering the lift and allowing the bird to stabilize itself or to maneuver and turn.

COMPARATIVE ANATOMY

Crushing and tearing

An eagle tears animal flesh with its sharp-edged bill, but it does not need to crush food in the way that some birds do. The muscles of an eagle's jaws are not as powerful as those of a grosbeak, for example, which is able to break open strong seeds and fruit stones simply by applying great force through its jaws.

expand the secondary feathers in other birds. The muscles in front of the radius and ulna (together referred to as the forearm) extend the flight feathers, while those behind the forearm close the flight feathers.

The forearm muscles are also used to twist the outer wing, a movement vital for powered flight to take place. The patagialis brevis and patagialis accessorius muscles stiffen the wings once the bird is in flight.

The leg muscles

While the muscles of a bird's wing are unique, there are many similarities between birds' leg muscles and those of mammals and reptiles. Leg muscles are very important for eagles. They act as shock absorbers when the bird lands at speed, allow prey to be grasped and

killed, and enable the eagle to walk or run on the ground. The tarsus (lower leg) is a channel for tendons from muscles in the upper leg that operate the toes and claws. In larger eagles these muscles extend into the lower leg, adding to the awesome power of an eagle's foot. The sartorius muscles bend the legs.

Head and tail muscles

An eagle needs to scan a wide field of view in search of prey. Since its eyes cannot rotate in their sockets, the eagle has to turn its head to scan. Turning the head involves twisting and extending its long neck, and those movements, in turn, require well-developed muscles. Two bands of muscles in front of and behind the spine flex the neck forward and down and upward and back. The multifidis cervicis and intertransversales move the neck from side to side.

An eagle's tail muscles are important because the feathers of the tail, or rectrices, perform many functions. At different times the tail feathers act as a rudder, an air brake, and an elevator. The levator coccygeus and levator caudae muscles bend the eagle's tail upward. The depressor caudae muscles bend it downward, and the lateralis caudae muscles allow the bird to twist its tail and spread it into a fan shape. Fanning out the tail gives the bird more lift; twisting it helps the bird change direction when flying or gliding.

Nervous system

▶ **BRAIN AND SPINE**
Bald eagle

Messages are transmitted through the nervous system, with information coordinated by the brain and spinal cord.

brain
— cerebrum
— cerebellum
— optic lobe

spinal cord

brachial plexus

lumbosacral plexus

lumbosacral enlargement

pudendal plexus

pygidial nerves

Sensory neurons *bring information from sense organs for interpretation by the brain.*

spinal nerves

brachial nerves

Motor neurons *take information to muscles and other tissues to trigger a response.*

obturator nerve

sciatic nerve

There are two main parts of an eagle's nervous system: a central nervous system (CNS) consisting of the brain and the spinal cord and a peripheral nervous system (PNS) containing cranial and spinal nerves. As in other vertebrates, the fundamental units of a bird's nervous system are neurons, cells that conduct electrical impulses to and from the brain. The CNS receives impulses from the outside world and from within the body. After the information is processed, motor signals are sent through the spinal cord and PNS to control the bird's movements and reactions.

Into the brain

The avian brain is similar in many ways to that of a reptile, although it is larger relative to body size. Birds do not depend on a complex cerebral cortex for learning and memory. The three lobes of a bird's brain are the cerebrum, the cerebellum, and the optic lobe. The cerebrum is used in coordinating muscular activities concerned with locomotion and homeostasis (body equilibrium) and is mainly occupied by the corpus striatum. The optic lobe is large in all birds and especially in eagles. Eagles are very sensitive to sound and less sensitive to smell; it is their vision that is most acutely developed.

CLOSE-UP

Eagle eyes

Eagles soar up to 5,000 feet (1,500 m) above the ground. At such an altitude they can scan a wide range of ground and sky, but they can spot prey only because they have very sharp eyesight. In fact, eagle eyes (right) are among the sharpest of any vertebrate. Eagles have forward-facing binocular vision, giving them excellent distance judgment. Their eyesight gives eagles a huge advantage when hunting wary prey. An eagle's eyes are large relative to the skull; a bald eagle has bigger eyes than a human. This allows the maximum amount of light to be captured. The retina contains an area called the fovea centralis, packed with light-sensitive photoreceptor cells, or rods and cones. This is where the eye's acuity is at its sharpest. In a human eye, 200,000 photoreceptors occur in the fovea. In the eye of a common buzzard, another member of the Acciptridae, it contains more than 1 million cones alone; a buzzard's eyesight is seven times sharper than that of a human. Also, an eagle sees an image about 30 percent larger than a human does.

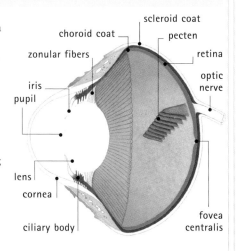

choroid coat
zonular fibers
iris
pupil
lens
cornea
ciliary body
sscleroid coat
pecten
retina
optic nerve
fovea centralis

Digestive and excretory systems

Like all birds, eagles have a faster and more efficient digestive system than other vertebrates. Eagles can digest food very quickly. Their diet varies greatly. Bald eagles eat mostly fish; some eagles prey almost exclusively on snakes and other small reptiles; others consume mammals and birds. Despite this variation, the key organs of an eagle's digestive and excretory systems remain similar to those of other birds.

The nature of the bill determines which foods birds eat. Eagles may kill prey that is almost as large as themselves, and is certainly too large to fit into the mouth and esophagus. The sharp cutting edges of an eagle's bill allow it to tear off small chunks of meat.

Useful crop

The esophagus is a long, thin tube connecting mouth to stomach. Like other birds, eagles have a pouch partway along the esophagus called a crop, inside which food is stored. This is a handy way to feed young birds, since food is rarely caught close to the nest. An eagle will return with a full crop, then regurgitate the contents into the mouth of the chick.

Eagles have a two-part stomach. The first section is a tube called a proventriculus. The walls of this part of the stomach emit mucus, acids, and peptic enzymes (chemicals that break down proteins) to start the digestion process. Food then enters the second part of the stomach, the gizzard. The gizzard performs a role similar to the molar teeth of mammals,

---

IN FOCUS

Regurgitation

Eagles do not attempt to digest all the food they swallow. They cough up, or regurgitate, indigestible matter, such as bones, hair, fish scales, and feathers in the form of pellets. Close examination of the contents of pellets can give many clues to the birds' diet.

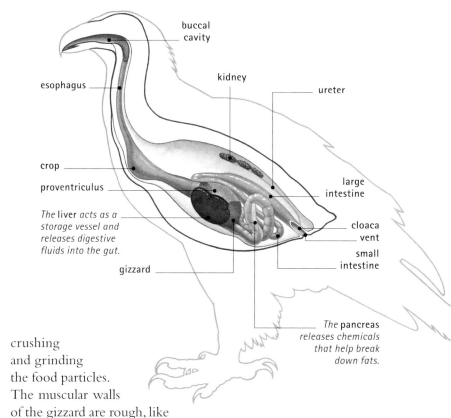

crushing and grinding the food particles. The muscular walls of the gizzard are rough, like sandpaper. An eagle's gizzard can crush even large bones, though most are regurgitated.

Food is digested and absorbed in the small intestine. Eagles have a short small intestine compared with birds that eat plant material. The walls of a bird's small intestine have pouches called intestinal ceca. These are small in eagles but contain bacteria that help break down food. Nutrients are absorbed through the walls of the intestines and pass into the bloodstream, mainly into the vessels of the superior and inferior mesenteric veins.

Nutrient storage and transport

The nutrients travel into the portal hepatic blood system inside the liver. There they are processed and stored before being distributed to the body tissues. Birds have a much larger liver than mammals. A large liver offers rapid storage and retrieval of nutrients.

Birds have a relatively short large intestine connecting the small intestine to the cloaca. Wastes are stored at the cloaca before they are expelled from the vent. Birds, unlike mammals, deposit solid and liquid wastes together.

▲ **GUTS AND KIDNEYS Bald eagle**
Birds have a pair of three-lobed kidneys. These filter waste materials from the blood. The wastes contain little water; they pass along a pair of tubes called ureters that exit the kidneys midway between the front and middle lobes. The ureters open into the top of the cloaca.

Circulatory and respiratory systems

CONNECTIONS

COMPARE the air sacs of an eagle with the trachea of an insect such as a *DRAGONFLY*.

COMPARE the heart of an eagle with the massive heart of a *GIRAFFE*.

Oxygen is carried around an eagle's body in its bloodstream. As with mammals, the heart pumps oxygenated blood around the body in arteries and the pulmonary vein, and deoxygenated blood is pumped through veins and the pulmonary artery. In the lungs, the blood releases waste carbon dioxide gas before picking up more oxygen. Some of an eagle's arteries and veins lie next to each other. This proximity allows warm blood leaving the heart and flowing around the body to heat cooler blood returning to the heart. In this system, a countercurrent heat exchange, some heat is constantly returned to the body core and helps eagles that live in cold climates conserve body heat.

Learn about lungs

To sustain active flight, with many wing beats each minute, an eagle's muscles require a constant supply of oxygen. This in turn requires an extremely efficient blood circulation system. Relative to body size, the hearts of birds are much bigger than those of mammals. An eagle's lungs typically occupy only 2 percent of the body volume. However, birds also possess a complex series of air sacs. They occupy around 15 percent of the body volume. In mammals and reptiles air moves in and out of the lungs in a tidal flow pattern. Yet in eagles and other birds, the combination of lungs and air sacs, together with certain hollow bones, allows a continuous stream of air to pass through the lungs in a one-way flow.

A continuous flow through the lungs provides a constant supply of oxygen. This allows eagles to remain active at very high altitudes and for long periods. Eagles can fly very high, and some can accomplish long migration flights. One relative of the eagle, Rüppell's griffon vulture, can fly at up to 35,000 feet (11,000 m) in the Himalayas.

The path of air

Air travels in through the mouth, along the trachea, through the lungs, through the system of air sacs and hollow bones, and out of the trachea again. The air sacs help ensure that the maximum amount of air passes through the lungs, but since they are poorly supplied with blood vessels they do not contribute directly to gas exchange, which takes place only in the lungs.

Keeping cool with air sacs

As well as allowing a continual stream of oxygen-rich air to pass through the lungs, the air sacs serve another important function. The evaporation of moisture on their inner surfaces helps reduce the eagle's temperature. Eagles generate a lot of heat as they fly, and they are covered with feathers, which are very good insulators and do not allow much heat to escape. Air sacs provide a great way to reduce body temperature.

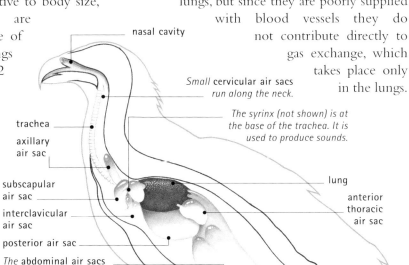

nasal cavity

Small cervicular air sacs *run along the neck.*

The syrinx (not shown) is at the base of the trachea. It is used to produce sounds.

trachea

axillary air sac

subscapular air sac

interclavicular air sac

posterior air sac

lung

anterior thoracic air sac

The abdominal air sacs *act as the main air reservoir. These are connected to the hollow long bones.*

▲ AIR SACS AND LUNGS
Bald eagle
As well as the lungs and air sacs, the long bones (not shown) are important for respiration in eagles.

Reproductive system

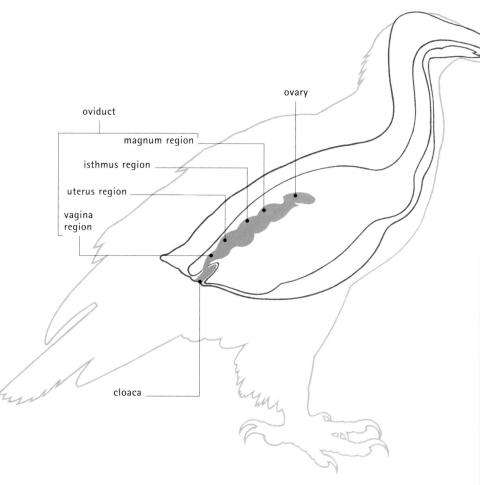

oviduct
magnum region
isthmus region
uterus region
vagina region
ovary
cloaca

▲ **FEMALE REPRODUCTIVE SYSTEM**
Unlike most other birds, eagles have two functioning ovaries rather than only one. This is a feature that eagles share with other raptors.

A male eagle's reproductive system produces sperm (male sex cells) in the testes; females produce ova or eggs (female sex cells) in the ovaries. When a pair of eagles copulates, sperm passes from the male's testes along the deferent duct into the cloaca. Unlike some birds, eagles do not have a penis, so the cloacae of the male and the female are brought closely together. This allows the sperm to pass across. The sperm then travels up the oviduct to a region near the ovary called the ostium.

Inside the female, an ovum passes from the ovary to be fertilized at the ostium. The fertilized egg is moved by cilia (tiny threadlike filaments) on the walls of the oviduct to the magnum region. There it is coated with albumen, or "egg white." Then the egg is carried by cilia to the isthmus region

COMPARATIVE ANATOMY

Two ovaries

Eagles are unusual among birds in that both ovaries and oviducts of the females are functional. In most birds besides eagles and other diurnal (day-flying) raptors, just the left ovary and oviduct function.

▼ *The path of an egg through a female eagle's reproductive system. The section between the ovary and the cloaca is called the oviduct.*

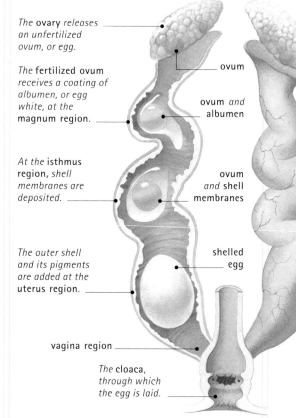

The **ovary** *releases an unfertilized ovum, or egg.*

The **fertilized ovum** *receives a coating of albumen, or egg white, at the magnum region.*

At the **isthmus** *region, shell membranes are deposited.*

The outer shell and its pigments are added at the uterus region.

ovum

ovum *and* albumen

ovum *and* shell membranes

shelled egg

vagina region

The **cloaca**, *through which the egg is laid.*

▲ *A bald eagle with its chick. A female eagle lays a clutch of one or two (sometimes three) eggs each year. Birds such as eagles that have a long life expectancy usually have smaller clutches than those that tend to die younger. A large eagle may live to 40 years and produce 50–60 eggs in its lifetime. A small songbird, with a life expectancy of 2 or 3 years, may lay clutches of 15 eggs each year.*

of the oviduct. A protein called keratin is deposited to form a hard, tough coating over the egg. This coating forms the eggshell. On the next stage of its journey, the egg passes into the uterus region of the oviduct. There, pigment is added to the shell. This gives it a distinctive coloring—no two species of birds have identical egg patterns and colors.

Finally, the egg is moved to the cloaca before being laid. The embryo develops in the central part of the egg, or yolk. Oxygen diffuses (moves) in through pores in the shell. Blood takes the oxygen to the embryo and carries carbon dioxide in the opposite direction. Other wastes accumulate in the shell. After 45–60 days an eagle embryo is sufficiently formed to break out of its egg. The hatched chick will then be cared for by its mother or both parents until it is able to look after itself.

TIM HARRIS

COMPARATIVE ANATOMY

Egg size and hatching times

A bald eagle egg is 2.5 to 3 inches (6 to 8 cm) long and weighs between 4.3 and 4.6 ounces (120 and 130 g). The yolk provides enough food to nourish the growing embryo until it is strong enough to hatch, after 34 to 37 days of development. By contrast, the eggs of a vervain hummingbird, at 0.4 inch (1 cm) long and just 0.013 ounce (0.375 g) in weight, support the embryo for only about 11 days before hatching occurs.

FURTHER READING AND RESEARCH

Parry-Jones, J. and F. Greenaway. 2000. *Eyewitness: Eagles and Birds of Prey.* Dorling Kindersley: London.

Proctor, N. S. and Patrick J. Lynch. 1993. *Manual of Ornithology.* Yale University Press: New Haven, CT.

Stone, Lynn, M. 2003. *Bald Eagles.* Lerner Publishing Group: Minneapolis, MN.

CLOSE-UP

Reproductive hormones

Hormones are messenger chemicals that tell cells what to do. They are secreted into the bloodstream from a series of glands that form the endocrine system. Major endocrine glands in birds include the liver and pancreas, as well as the pituitary, thyroid, parathyroid, and adrenal glands. The liver and pancreas are also part of a different system, the exocrine system; some of the secretions they produce are released onto the surface of the gut cavity.

Some hormones are vital to the reproductive process. The parathyroid gland produces a hormone that helps regulate levels of phosphorus and calcium in the body; this is critical for female birds in the breeding season, since calcium is needed for eggshell production. The male's testes are endocrine glands—they produce a hormone called testosterone—and the female's ovaries secrete estrogen and testosterone. The adrenal gland produces sex hormones in birds of both sexes.

Earthworm

PHYLUM: Annelida CLASS: Clitellata: ORDER: Oligochaeta
FAMILIES: Lumbricidae and several others

Earthworms belong to the Oligochaeta, a large group of about 3,500 species. Not all oligochaetes are earthworms, though. Most oligochaetes live in freshwater pools and streams; earthworms are terrestrial oligochaetes. Since they occur in several families, there is no one earthworm group.

Anatomy and taxonomy

Biologists place organisms into taxonomic groups according to their anatomy, although studies of DNA are an increasingly important tool. The tree below shows the classification of one of the most familiar earthworms, the common earthworm, or night crawler. Earthworms are annelids, or segmented worms. This major group also includes the polychaete worms, a marine group.

● **Animals** These are multicellular organisms that have eukaryotic (nucleus-containing) cells without cell walls. These cells are usually organized into tissues and organs. Animals rely on other organisms for energy; almost all have guts inside which digestion takes place. Most animals have muscles to allow movement and a nervous system that enables them to interact with their surroundings.

● **Annelid worms** The annelid or segmented worms are a large and diverse group. Annelid bodies are built up from large numbers of similar segments, arranged one behind the other. Annelids are metameric—internal structures tend to repeat again and again along the body. All annelids have three main body regions. They are the head, which is further divided into a peristomium, which surrounds the mouth, and a prostomium, which overhangs it; the trunk, which contains repeating segments bearing bristlelike setae; and the region at the end of the body, the pygidium.

Annelid internal anatomy can be described as a tube within a tube. The inner tube is the gut, and the outer tube is the outer body wall. The space running in between is the body cavity, or coelom. The coelom is usually divided by septa, walls that separate each segment from its neighbors.

● **Polychaete worms** The largest annelid group is the polychaetes. An almost exclusively marine group, polychaetes have lobes called parapodia running in pairs along their bodies. These lobes are used for gas exchange and for movement; the setae that project from the lobes also help the animals move about. Polychaetes bear tentacles on the head.

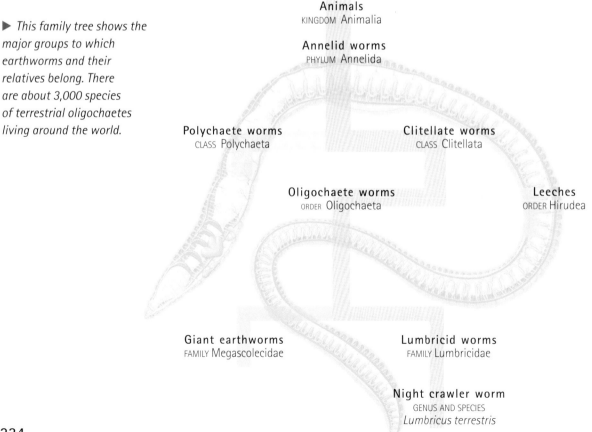

▶ This family tree shows the major groups to which earthworms and their relatives belong. There are about 3,000 species of terrestrial oligochaetes living around the world.

Animals
KINGDOM Animalia

Annelid worms
PHYLUM Annelida

Polychaete worms
CLASS Polychaeta

Clitellate worms
CLASS Clitellata

Oligochaete worms
ORDER Oligochaeta

Leeches
ORDER Hirudea

Giant earthworms
FAMILY Megascolecidae

Lumbricid worms
FAMILY Lumbricidae

Night crawler worm
GENUS AND SPECIES
Lumbricus terrestris

• **Clitellate worms** Worms from this class mainly live in freshwater or on land. Clitellates are characterized by the clitellum or saddle, a broad band of skin about one-third of the way along the animal's body. The saddle secretes a cocoon inside which eggs develop. Clitellates are hermaphrodites—they contain both male and female reproductive structures. Their young (unlike the young of polychaetes) do not have a larval stage. Polychaete larvae are different from the adults and swim among the plankton, but clitellate young emerge from their eggs as miniature adults.

• **Leeches** One of the main clitellate groups is the leeches. Most leeches are parasites that attach to and steal blood from larger animals. They have a large sucker at each end of the body, which they use to aid locomotion or to anchor to a host. Leeches have rasping mouthparts to bite into flesh, and release chemicals called anticoagulants to keep the blood flowing. Leeches do not have setae, and internally there is little evidence of metamerism (the repeating of organs); the septa are absent and much of the body cavity is filled by tissue that contains fluid-filled channels. In leeches, unlike oligochaetes, fertilization occurs inside the body; fertilized eggs are laid in a cocoon.

• **Oligochaete worms** Earthworms and their relatives form the order Oligochaeta. Oligochaetes do not have parapodia as polychaetes do—instead, they breathe through their skin. They have rows of setae on their outer surface (though fewer and smaller than in polychaetes). Setae are used for burrowing and for attaching to a mate. Unlike leeches, oligochaetes can regenerate lost body parts. They are segmented, so muscles can contract or relax in sequence to power locomotion through water or dirt.

• **Earthworms** Earthworms occur in several different families. In North America, most belong to the families

▲ *This polychaete filters sediment from seawater with its fronds.*

Megascolecidae and Lumbricidae. Biologists use features such as length, distance between head and saddle, and color to identify earthworms. Recognition to species level depends on the arrangement of bumps that surround the gonopores (pores through which sex cells leave the body). Segment number is read from the front of the animal; the male gonopores are normally on the 15th segment.

• **Night crawler** *Lumbricus terrestris*, the night crawler, is one of the most familiar earthworms. It is the species you are likely to see on a sidewalk after heavy rains or when digging in your backyard. This reddish-brown worm grows up to 10 inches (25 cm) long and digs burrows up to 8 feet (2.5 m) deep.

FEATURED SYSTEMS

EXTERNAL ANATOMY An earthworm is divided into many segments separated by grooves. A thin, moist cuticle penetrated by various pores covers each of the earthworm's segments. *See pages 236–237.*

INTERNAL ANATOMY Earthworms' internal anatomy is dominated by metamerism, the repeating of organs, though the degree of metamerism varies greatly between systems. *See pages 238–239.*

NERVOUS SYSTEM Earthworms have a tiny brain connected to a nerve cord that stretches the length of the body. Giant neurons help messages pass to the brain rapidly when the animal is threatened. *See page 240.*

CIRCULATORY AND RESPIRATORY SYSTEMS Earthworms have a closed circulatory system; that is, the blood stays within blood vessels. Two major vessels are connected by five pairs of beating hearts. *See page 241.*

DIGESTIVE AND EXCRETORY SYSTEMS Organic material is stored and ground before entering the midgut for digestion. Waste is filtered from body cavity fluids by specialized organs called nephridia. *See page 242.*

REPRODUCTIVE SYSTEM Earthworms have both male and female reproductive organs, with the saddle producing a cocoon inside which the eggs are fertilized and go on to develop. *See page 243.*

External anatomy

CONNECTIONS

COMPARE the body plan of an earthworm with a snake such as a *GREEN ANACONDA.*

COMPARE a worm's skin with that of an insect such as an *ANT.*

At first glance, an earthworm seems to have no external features. But look closely and you will see that it is composed of many segments separated by tiny grooves. Covering each segment is a thin, tough cuticle. This is formed by interwoven fibers of a tough protein called collagen. Directly under the cuticle is a layer of cells called the epidermis, from which the cuticle is secreted.

Immediately in front of the first body segment at the head of the animal are two small unsegmented regions. The peristomium circles the mouth; on its dorsal (top) surface is a small notch into which a section

COMPARATIVE ANATOMY

Mouthpart diversity

Annelid lifestyle has a major influence on anatomy. This can be illustrated by comparing annelid mouthparts. Earthworms use their peristomium to funnel food into their mouth. Earthworms do not need teeth, since their food is ground up in the gizzard, but other annelids do. A leech's mouth opens inside a sucker on its head with which it attaches to hosts. Leeches have three sets of sharp teeth. They use these to bite a Y-shaped hole in their victim's body. They then drink the host's blood. Some predatory polychaetes have a structure called a proboscis in the head at the top of the gut. This can be everted (popped out) rapidly, and bears powerful jaws for grasping prey.

called the prostomium fits. The peristomium acts as a kind of lip. The earthworm uses its peristomium to grasp and manipulate its food.

Saddle and skin

Earthworms have no eyes or ears, but farther along the body from the head is a band of thickened skin called the saddle. This develops when the earthworm becomes sexually mature. It is important for producing mucus and forming an egg case. Segments continue for the length of the body. Extending from the last body segment is a structure called the pygidium. The pygidium contains an opening called the anus through which feces pass.

Super setae

Each segment bears four pairs of tiny bristles called setae. Made of a hard material called chitin, they emerge from the cuticle through pores and are controlled by tiny muscles. Two

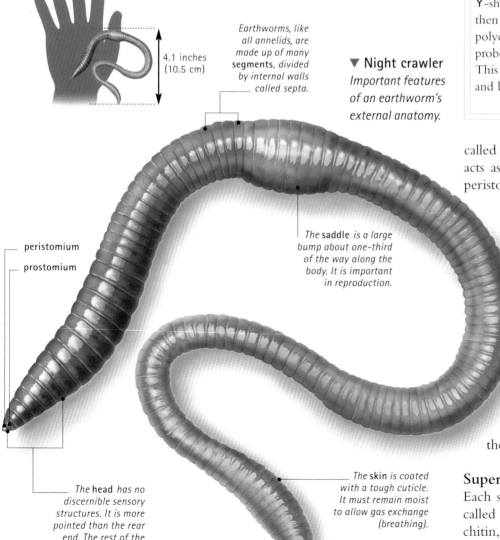

4.1 inches (10.5 cm)

Earthworms, like all annelids, are made up of many segments, divided by internal walls called septa.

▼ **Night crawler**
Important features of an earthworm's external anatomy.

The **saddle** *is a large bump about one-third of the way along the body. It is important in reproduction.*

peristomium

prostomium

The **head** *has no discernible sensory structures. It is more pointed than the rear end. The rest of the body forms the* **trunk.**

pygidium

The **skin** *is coated with a tough cuticle. It must remain moist to allow gas exchange (breathing).*

pairs of setae extend laterally (from the side); the other two pairs are on the ventral (lower) side of each segment.

Setae are essential for earthworms to get about. Used in combination with bands of muscle inside the body, they can be extended to wedge a segment into the soil. They can then be withdrawn to allow the segment to slide forward. Setae are powered by their own set of tiny muscles. As well as movement, they are important for binding to mates.

Bumps, pores, and holes

An earthworm's cuticle has many bumps and holes. Wastes processed by excretory organs called nephridia empty through pores in the cuticle. These pores are opened and closed by rings of muscle. Other holes, called gonopores, provide exits for sex cells—sperm and eggs. On the ventral surface of the 15th segment is a set of bumps. They surround the male gonopores, through which sperm leave the body. Eggs leave the body through a different pair of gonopores, on the 14th segment. Nearer the head are pores connected to two pairs of sacs called spermathecae. They store sperm received from another earthworm.

Earthworms need a moist cuticle to breathe. They release fluid from the body cavity, or coelom, through tiny pores in the grooves that separate the earthworm's body segments. Muscles keep these coelomic pores closed for most of the time. This helps the animal reduce unnecessary water loss. Mucus is also released onto the cuticle through pores connected to cells in the epidermis.

▼ *The head of an earthworm. The skin of the animal is lubricated with mucus and with fluid released from the coelom. This allows the worm to breathe through its skin.*

Internal anatomy

Metamerism, the repeating of structures in segments along an animal's body, is less pronounced in earthworms than in polychaetes. Polychaete organs tend to occur repeatedly through the body. Most earthworm organs are located in just a few segments. Nevertheless, metamerism is apparent to a degree in most earthworm body systems. For example, the ventral nerve cord runs along the length of the body, with swellings called ganglia in each segment, each with three pairs of nerves splitting off—metameric features. However, at the front of the body the nerve cord splits, circling the pharynx (top of the gut) to merge with the brain. Nerves that leave the nerve cord in this region show little sign of metamerism.

Degrees of repetition

Two long blood vessels run the length of an earthworm's body. Smaller tubes branch from the vessels in each segment; one extends to the body wall, the other to the gut. Near the head, five pairs of hearts pump the blood from one vessel to the other. One of the few truly metameric systems is the excretory system. A pair of excretory organs, the nephridia, occurs in almost all the segments of the body.

Little metamerism is apparent in the digestive or reproductive system. The gut runs through the center of the earthworm, extending from the mouth through a number of sections such as the crop and the gizzard. The gut ends at the anus, through which material that cannot be digested passes.

Earthworms have two pairs of testes (sperm-producing organs) that lie in the 10th and 11th segments. A single pair of ovaries (egg-producing structures) occurs in the 13th body segment.

There are two main types of muscles inside an earthworm. A series of muscles runs around the worm, just beneath the skin. They are called the circular muscles. Inside them lies a set of longitudinal muscles. They run along the body from the head to the pygidium (final segment). Each longitudinal muscle links three or four separate segments together.

Bounding the coelom

The inner surfaces of the longitudinal muscles are covered by a thin layer of cells called the peritoneum. The peritoneum also coats

intestine

dorsal blood vessel

gizzard

The testes lie near the base of the seminal vesicle.

esophagus

hearts

pharynx

brain

subpharyngeal ganglion

spermathecae

ovary

ovisac

oviduct

crop

ventral blood vessel

segmental blood vessels

segmental nerves

*The **nephridia** are excretory structures. Note that they extend to just behind the pharynx, but those near the head have been omitted from this illustration.*

septa

anus

▲ Night crawler
Important features of a night crawler earthworm's internal anatomy. Note the metameric nature of the excretory and circulatory systems.

One becomes two

Have you ever chopped an earthworm in half accidentally while digging in the backyard? Do not feel too sad for the worm the next time you do, because new life rather than death may well be the outcome. Worms have amazing powers of regeneration. The wound on each half soon heals. Next, a ball of cells called a blastema develops at the site of the wound. The cells that form the blastema are undifferentiated—they are able to develop into any other type of earthworm cell. Each contains a copy of the original worm's inherited material, or DNA. The cells are able to use the instructions encoded by the DNA to rebuild the lost body part. Chopped worms can grow a new head (including a new brain) or a new hind region. So, in a short amount of time one chopped worm can become two new ones.

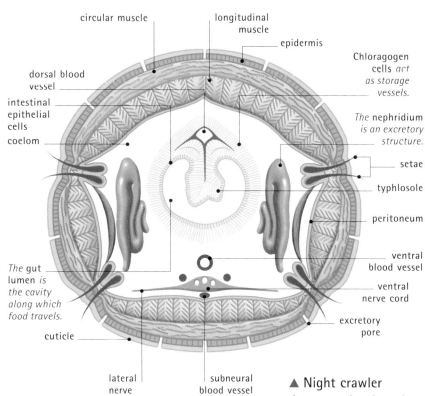

▲ **Night crawler**
A cross section through an earthworm showing important features of internal anatomy.

▼ *These earthworms have escaped the chop, but if they had been sliced in two they could have regenerated missing body parts.*

the surfaces of the internal organs, so it acts as a boundary for the body cavity, or coelom. The individual segments of an earthworm's body are divided internally by wall-like structures called septa. The septa divide up the coelom, which lies between the gut and the muscular layers. Each segment effectively has its own coelom.

The hydrostatic skeleton
Vertebrate muscles attach to bones. Earthworms are invertebrates, animals that do not have bones against which muscles can pull. Invertebrate muscles may instead be braced against the fluid-filled coelom. This arrangement is called a hydrostatic skeleton. A hydrostatic skeleton works because of the incompressibility of fluids: the fluid inside the coelom cannot be squashed. This resistance to compression provides a brace against which the muscles can operate. Hydrostatic skeletons are very common in invertebrates.

To move forward, an earthworm first contracts the circular muscles near the head. The contraction makes the segments stretch, while the hairlike setae retract

to allow easy passage through the soil. A wave of circular muscle contractions passes down the body. This is soon followed by a wave of longitudinal muscle contractions that makes each segment bulge outward in turn. Together with the extension of the setae, this bulging wedges the segment into the soil, allowing the animal to draw itself forward.

Nervous system

Earthworms need a nervous system to interpret sensory information from their surroundings and to coordinate a response. At the hub of the nervous system is the brain. This lies in the prostomium, where it rests just above the pharynx (part of the digestive system). The brain consists of two pear-shaped lobes, connected by a narrow stretch of tissue called a commisure. A pair of large nerves extends from the brain and loops around the pharynx.

Beneath the pharynx, the nerves merge to form a nodule of nervous tissue called the subpharyngeal ganglion. Nerves extending forward from this ganglion control the sections that make up the head of the animal. These include the peristomium, the lip that the animal uses to draw food into the mouth.

Connecting brain and body

A long bundle of nerve fibers called the ventral nerve cord extends from the subpharyngeal ganglion. Surrounded by a fibrous sheath, the nerve cord runs along the underside of the worm's body. Swellings occur in the cord as it passes through each segment. They are called segmental ganglia. From each ganglion three pairs of nerves extend into nearby tissues and muscles. Each ganglion controls most of the physiological functions of each segment. They also help coordinate movement of the animal.

Short on sense organs

Unlike polychaetes, many of which have eyes, palps, and long tentacles to feel around, earthworms do not have specialized sensory structures. Since earthworms spend most of their life underground, they have little need for eyes. However, they do have light-sensitive cells that lie just under the skin. These are concentrated at the front and rear of the animal, and on the dorsal (top) surface.

The skin also contains cells along its length that detect chemicals, and cells called mechanoreceptors that are connected to hairs. Mechanoreceptors detect movements of the hairs and translate them into nervous signals. These sense cells enable earthworms to detect vibrations passing through the soil, helping them avoid burrowing predators such as moles.

Speeding up escape

Earthworms respond to attack with a sudden shortening of the body. This is important for earthworms such as night crawlers, which venture out of their burrows to feed; night crawlers drag their food of decaying leaves underground. When they are threatened, a relay of nervous messages from segment to segment would take too long and might lead to death rather than escape. Instead, earthworms have three giant nerve fibers running along the nerve cord. These do not connect to other nerves, so signals can travel the length of the body very quickly.

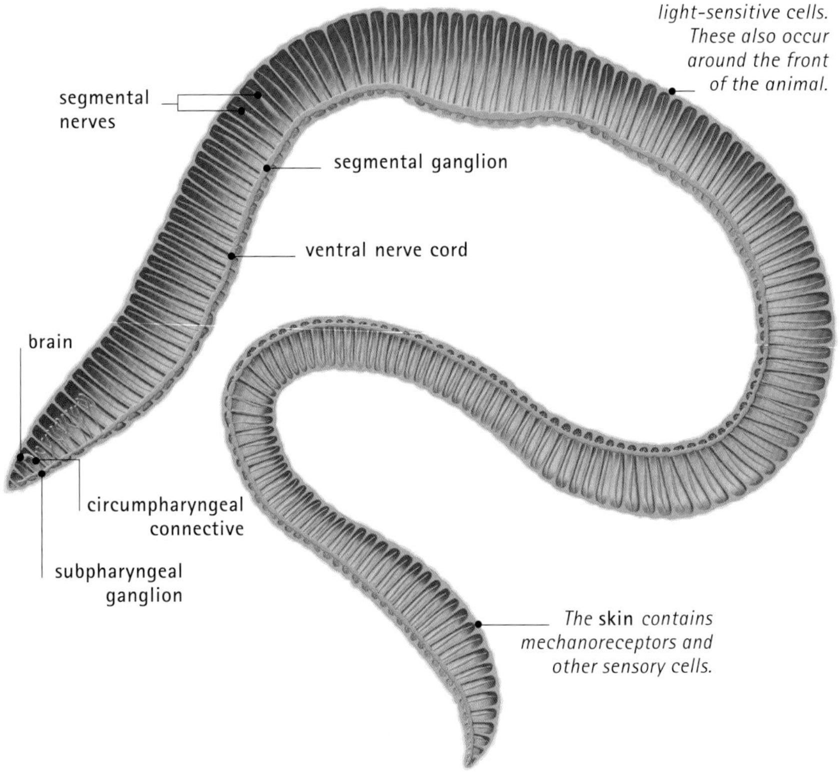

▼ **Night crawler**
Features of the earthworm nervous system. Each segmental ganglion gives rise to three main nerves (one is shown here); these then divide further.

*The **dorsal surface** of the skin contains light-sensitive cells. These also occur around the front of the animal.*

segmental nerves

segmental ganglion

ventral nerve cord

brain

circumpharyngeal connective

subpharyngeal ganglion

*The **skin** contains mechanoreceptors and other sensory cells.*

240

Circulatory and respiratory systems

CONNECTIONS

COMPARE the closed circulatory system of an earthworm with the open system of an insect such as a *HOUSEFLY*.

COMPARE the respiratory pigment of an earthworm with that of a *CRAB* or an *OCTOPUS*.

Many polychaetes have gills, outgrowths of the body wall that take oxygen from the water around them. Most oligochaetes do not have gills. Instead, they draw the oxygen they need through their skin. Both strategies rely on diffusion, the movement of molecules from areas of high concentration to lower concentrations. An earthworm's skin must be moist for diffusion to take place, so it releases mucus and other fluids onto its cuticle. Oxygen dissolves into the fluid and then moves into the body. Waste carbon dioxide gas moves the other way through the moist skin.

The passage of blood

Once inside the body, oxygen is carried by the earthworm's blood to wherever it is needed. Earthworms have a closed circulatory system; that is, blood is contained within vessels. The blood carries nutrients as well as oxygen to cells and takes their wastes away. Two major blood vessels run through the body. The dorsal blood vessel lies just above the gut. Circular rings of muscle in the vessel walls contract to force blood forward, with valves stopping the blood from flowing back. The blood then reaches five pairs of hearts that lie in consecutive segments near the head. The hearts force the blood into the ventral vessel, which runs along the underside of the gut.

Blood moves back along the body, leaving via smaller vessels that branch off in each segment. The largest of these is the subneural vessel, which supplies the ventral nerve cord. After traveling to the organs and muscles, the blood returns to the dorsal blood vessel.

Pigments without cells

Both earthworm blood and human blood are red. That is because both contain a pigment called hemoglobin. Hemoglobin binds to oxygen and carries it to cells that need it. Human hemoglobin is contained within cells, but earthworm hemoglobin is free, forming massive molecules that dissolve in the fluid that makes up most of the blood.

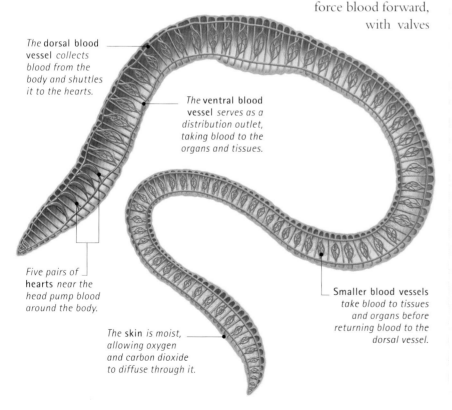

▼ **Night crawler**
The skin provides for all the animal's gaseous exchange needs.

The **dorsal blood vessel** *collects blood from the body and shuttles it to the hearts.*

The **ventral blood vessel** *serves as a distribution outlet, taking blood to the organs and tissues.*

Five pairs of hearts *near the head pump blood around the body.*

The **skin** *is moist, allowing oxygen and carbon dioxide to diffuse through it.*

Smaller blood vessels *take blood to tissues and organs before returning blood to the dorsal vessel.*

COMPARATIVE ANATOMY

Opposite arrangements

The main annelid blood vessels run just above and below the gut, with the nerve cord running ventrally to (underneath) the lower one. This is opposite to the arrangement in chordates, the group that includes vertebrates. Such a fundamental difference in body plan suggests that the common ancestor of the two groups lived a very long time ago. Some biologists think that ancient ancestors of chordates had an annelid-like arrangement. They argue that at some stage the body of these ancestors became inverted (flipped upside-down). An alternative explanation is that annelid and chordate ancestors had blood and nervous systems spread throughout the body. After the two groups separated, the systems evolved nerve cords and blood vessels independently, with annelid systems evolving one way and chordates the other.

Digestive and excretory systems

Earthworms feast on decaying matter as they burrow through it. Food is pushed by the peristomium into the mouth, or buccal cavity, which extends from segments 1 through 3, and then into the pharynx. This is attached to muscles connected to the body wall that pull the pharynx open. Muscles in the pharynx push food into the next section, the esophagus, which extends from segments 6 through 12. At the rear of the esophagus are two pairs of calciferous glands; they secrete crystals of calcium carbonate that keep the pH (relative acidity) of the gut at just the right level.

Next, the food passes into a bulbous storage vessel called the crop, then into the gizzard. This muscular part of the foregut grinds food down into fine particles.

▲ *Worm casts are rich in nutrients and encourage plant growth.*

The top of the intestine is heavily folded to form a structure called a typhlosole, through which sugars and other nutrients pass into the body. The folds increase the surface area available for absorption. Lining the typhlosole is a bank of yellow chloragogen tissue. Chloragogen acts like a liver; it stores fat and glycogen (a starch) and breaks down proteins.

Hindgut and out

Scoured of nutrients, the remaining dirt passes through the hindgut, or rectum, before eventually being excreted at the anus. Worm feces, or casts, form an extraordinarily rich soil.

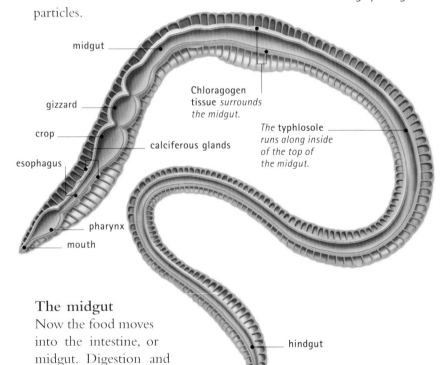

midgut

gizzard

crop

esophagus

Chloragogen tissue *surrounds the midgut.*

calciferous glands

The **typhlosole** *runs along inside of the top of the midgut.*

pharynx

mouth

hindgut

anus

The midgut

Now the food moves into the intestine, or midgut. Digestion and absorption of the food take place here. Unlike the rest of the digestive system, the intestine is not lined with tough cuticle. This allows easier uptake of nutrients. The front part of the intestine is important for making and releasing enzymes. These are chemicals that break down food into much smaller molecules that are easily absorbed into the body.

▲ **Night crawler**
The digestive tract of an earthworm. The pharynx is opened by muscles that attach to the body wall.

Getting rid of waste

Almost every earthworm segment contains a pair of structures called nephridia. They are excretory organs that get rid of waste products. Each nephridium has a funnel-shaped opening into the coelom called the nephrostome. This leads into a long, coiled tube that connects with a pore through which waste is excreted onto the body surface. Nephridia contain dense colonies of bacteria. They help recycle proteins in the waste fluid. Other nutrients as well as much of the water are also reabsorbed.

Reproductive system

Unlike polychaetes, which have separate sexes, earthworms are hermaphrodites: they have both male and female reproductive organs. When two worms mate, they press their ventral surfaces together so the saddle of one is in contact with the 10th segment of the other. The worms are held in place both by the setae and by liberal quantities of mucus. Each worm releases sperm from the male gonopores, which lie on the 15th segment. The sperm travels along grooves lined with tiny beating filaments called cilia. They carry the sperm to openings on the 9th and 10th segments of the other worm. These openings lead to sperm storage banks, or spermathecae.

Fertilizing and laying eggs

A few days later, the saddle of each worm secretes a cocoon made of mucus. The worm squeezes its muscles to move the cocoon forward along its body. When it reaches the 14th segment, eggs are released into it from the female gonopore. As the cocoon moves over the spermathecae, it releases sperm it previously received from the other worm. The sperm fertilizes (fuses with) the eggs.

The cocoon moves over the head of the worm before being deposited. It shrinks to form a capsule. The capsule contains many fertilized eggs, but only one or a few will grow and develop into the hatchling stage. Growing polychaetes have a swimming stage (called a trochophore) which looks very different from the adults. The trochophore allows them to disperse to new areas. Hatchling earthworms do not need this stage. Instead, they look like miniature versions of their parents.

What is going on inside?

Earthworm reproductive structures occur in just a few segments. Sperm is made by the testes, which lie in the 10th and 11th segments. Freshly produced sperm passes into a pair of structures called seminal vesicles, where it matures. A duct called the vas deferens carries the sperm from the seminal vesicles to the male gonopores.

The female reproductive system includes the spherical spermathecae. The ovaries are small structures attached to the walls of the 13th and 14th segments. They produce the eggs, which pass to a storage vessel, the ovisac. Connected to the ovisac is a tube called the oviduct, along which eggs pass to the female gonopore. They then enter the mucus cocoon.

Not all earthworms mate to reproduce. Some are asexual—they do not need a mate to produce young. For example, *Pristina leidyi* forms a groove midway along its body. The front half of the worm then grows a rear end in its middle, while the other half grows a head—but the worm remains complete. Eventually the new head and tail break apart to leave two separate worms.

JAMES MARTIN

FURTHER READING AND RESEARCH
Petersen, Christine. 2002. *Invertebrates*. Scholastic Library Publishing: Danbury, CT.

▲ *Bound by mucus secreted from their saddles, a pair of earthworms exchange sperm. Each worm's male gonopores open into grooves that carry sperm to its partner's spermathecae.*

▼ **Night crawler**
Important features of an earthworm's reproductive system, which incorporates both male and female structures.

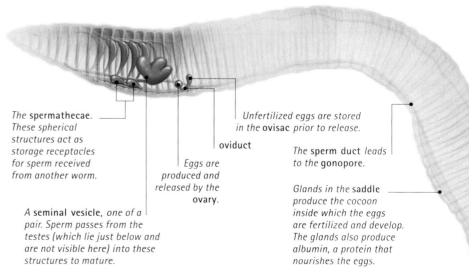

The **spermathecae.** These spherical structures act as storage receptacles for sperm received from another worm.

A **seminal vesicle,** one of a pair. Sperm passes from the testes (which lie just below and are not visible here) into these structures to mature.

Eggs are produced and released by the **ovary.**

oviduct

Unfertilized eggs are stored in the **ovisac** prior to release.

The **sperm** duct *leads to the* **gonopore.**

Glands in the **saddle** produce the cocoon inside which the eggs are fertilized and develop. The glands also produce albumin, a protein that nourishes the eggs.

Elephant

ORDER: Proboscidea FAMILY: Elephantidae
GENERA: *Loxodonta* and *Elephas*

The three species of elephants—the Asian elephant, African savanna elephant, and African forest elephant—are the sole remnants of the order Proboscidea. Until a few thousand years ago, however, there were many more species of proboscideans, including the mastodons, gomphotheres, and mammoths. Not all these beasts were giants: pygmy elephants lived on some Mediterranean islands until as recently as 4,400 years ago.

Anatomy and taxonomy

Proboscideans, the group that includes both elephants and extinct animals such as mammoths, share a number of unusual characteristics, such as a trunk and tusks, and are only distantly related to other mammals.

▶ *This family tree shows the major groups to which elephants belong, as well as some close relatives. After recent DNA analyses, the African elephants have been separated into two species. Note that the order Proboscidea contains many species that are extinct, including groups such as the mammoths and mastodons.*

- **Animals** All animals are multicellular and depend on other organisms for food. Unlike other multicellular organisms such as plants and fungi, most animals are able to move about and react quickly to stimuli.

- **Chordates** At some time in their life cycle, all chordates have a stiff supporting rod called a notochord running along the back of their body.

- **Vertebrates** The notochord of vertebrates is in the form of a spinal cord. This is encased by the spine, composed of bones called vertebrae. Most vertebrates are bilaterally symmetrical—the body shape is roughly the same on either side of the backbone. All vertebrates have a skull made of either bone or cartilage surrounding a brain.

- **Mammals** Mammals are unique among vertebrates in having mammary glands. Females nourish their young with milk secreted from these glands. Unlike all other vertebrates, most mammals have fur covering their body, a single lower jawbone that hinges directly with the skull, and red blood cells that lack nuclei.

Animals
KINGDOM Animalia

Vertebrates
SUBPHYLUM Vertebrata

Mammals
CLASS Mammalia

Placental mammals
SUBCLASS Eutheria

Proboscideans, sea cows, and hyraxes
SUPERORDER Paenungulata

Hyraxes
ORDER Hyracoidea

Elephants, mammoths, and mastodons
ORDER Proboscidea

Sea cows
ORDER Sirenia

Elephants and mammoths
FAMILY Elephantidae

African elephants
GENUS *Loxodonta*

Asian elephants
GENUS *Elephas*

African forest elephant
GENUS AND SPECIES
Loxodonta cyclotis

African savanna elephant
GENUS AND SPECIES
Loxodonta africana

Asian elephant
GENUS AND SPECIES
Elephas maximus

● **Placental mammals** These mammals develop inside the mother's uterus. There, they receive nourishment and oxygen through an organ called the placenta, which develops during pregnancy. The other major mammal groups are the egg-laying monotremes and the marsupials. They give birth to very small young that develop in a pouch on the surface of the mother's body.

● **Elephants, sea cows, and hyraxes** The orders that contain the elephants (Proboscidea) and the sea cows (Sirenia) are more closely related to each other than to other mammals. Sea cows are completely aquatic. Their front limbs have evolved into fins, and the rear limbs have disappeared altogether. They swim using large tail flukes. Dugong flukes resemble those of a whale, but manatees have a flat, round paddle. Hyraxes are also distant relatives of elephants. These small, rodentlike mammals live in rocky habitats or in trees in the Middle East and Africa.

● **Elephants and mammoths** Both living and extinct proboscideans have an elongated nose that forms a trunk and have a very large body. Like modern elephants, most extinct proboscideans bore tusks that protruded through the upper lip. Although modern elephants are the largest living land animals, some mammoths weighed more than twice the maximum recorded weight of an African savanna elephant. All living elephants are plant eaters, and so were their ancestors.

● **African elephants** The African savanna elephant is the biggest modern species, with large ears and a back that slopes down from the shoulders. Until recently, the African forest elephant was considered a subspecies (local form) of savanna elephant, but studies of forest elephant DNA have shown that it is genetically distinct. The forest elephant is

▲ *An African savanna elephant uses its trunk to pull vegetation from the ground and pass it to the mouth for chewing.*

smaller and darker than its savanna cousin and lives in the rain forests of west and central Africa. The savanna elephant lives on the grasslands of eastern and southern Africa. Both species have two fingerlike processes at the tip of the trunk.

● **Asian elephant** The Asian elephant has a humped back and relatively small ears, and the females lack tusks. The Asian elephant has just one fingerlike trunk process.

FEATURED SYSTEMS

EXTERNAL ANATOMY Elephants are massive, four-legged mammals. They have a large, short head with a trunk, large ears, and sometimes long tusks. *See pages 246–249.*

SKELETAL SYSTEM An elephant's skeleton is adapted for bearing great weight, with thick limb bones arranged like vertical pillars, so the weight passes along the length of the bones, reducing the need for muscular exertion and minimizing risk of injury to joints. *See pages 250–253.*

MUSCULAR SYSTEM Powerful leg muscles help the elephant maintain its posture when walking. The head, trunk, and neck are served by large muscles that support the trunk and tusks. *See pages 254–255.*

NERVOUS SYSTEM Elephants have the largest brain of any species of land animal and show a variety of complex behavioral adaptations, including use of tools. *See pages 256–257.*

CIRCULATORY AND RESPIRATORY SYSTEMS An elephant's oxygen requirement is less per unit of body weight than that of smaller mammals, and its heart rate is much slower. *See pages 258–259.*

DIGESTIVE SYSTEM Elephants eat large amounts of low-quality food. This is digested with the help of bacteria that live in the elephant's enlarged cecum at the junction of the small and large intestines. *See pages 260–261.*

REPRODUCTIVE SYSTEM Male elephants regularly enter a period of heightened aggression and sexual activity called musth. At such times, they seek out receptive females and fight off rivals. *See pages 262–263.*

External anatomy

The **ears** are very large. Richly supplied with blood vessels, they help the animal lose heat.

The **trunk** is the elongated nose and upper lip. A pair of nostrils run along its length.

Tusks are formed by the greatly enlarged upper incisors.

◄ African savanna elephar
Features of the extern
anatomy of an elephar

The **skin**
wrinkled an
very thick

The **legs** are wide and columnlike. They support the weight of the animal.

The **tail** can swish from side to side to deter flying insect pests.

13 feet (4 m)

17 feet (5 m)

COMPARE the elephant's tusks with the incisors of a *LION*, a *ZEBRA*, and a *HUMAN*.

COMPARE the posture of an elephant with that of a *RHINOCEROS* and a *HIPPOPOTAMUS*.

CONNECTIONS

Elephants have a huge but relatively short head, and a massive body supported by long, powerful, columnlike legs. Elephants have the longest nose, largest ears, and the longest teeth of any living mammal. In both African and Asian elephants, males weigh nearly twice as much as females. A large male Asian elephant weighs about 10,000 pounds (4,500 kg), and a good-sized male African savanna elephant weighs about 14,000 pounds (6,300 kg) and occasionally up to 16,500 pounds (7,500 kg).

African and Asian elephants have different body shapes. The middle of the back is the highest point of an Asian elephant, reaching up to 11 feet (3.3 m) high. The highest point of African savanna and forest elephants is the shoulders, rising up to 13 feet (4 m) in savanna males. Forest elephants are much smaller; the tallest are around 7 feet (2.2 m) tall.

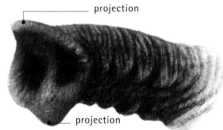

projection

projection

African elephant

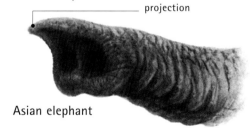

projection

Asian elephant

▲ TRUNK-TIP DIFFERENCES
Asian elephant trunks bear a single fingerlike projection at the tip. African elephant trunks have two of these projections.

The elephant's trunk

The most striking feature of an elephant is its trunk, which is a flexible, muscular, elongated nose. The trunk may be more than 6 feet (1.8 m) long in African savanna elephants. There are two nostrils at the tip of the trunk. Elephants also have fingerlike projections at the end of the trunk that they use to pick up small objects. African elephants have two projections, while Asian elephants have one.

Elephants use their trunk for both eating and drinking. The trunk is used to grasp hard-to-reach twigs or grasses close to the ground. Elephants also use their trunk to suck up water, which they squirt into their mouth for drinking or spray over their body to cool down. When walking through deep water or swimming, elephants can use their trunk as a snorkel to breathe even if submerged. Elephants are intelligent animals and can use their trunk to make tools, such as fly switches made from branches. Elephants even use their trunk to throw stones and other objects at animals that may be bothering them.

Giant incisors

All male elephants and female African elephants bear tusks—greatly elongated upper incisor teeth. Female Asian elephants have

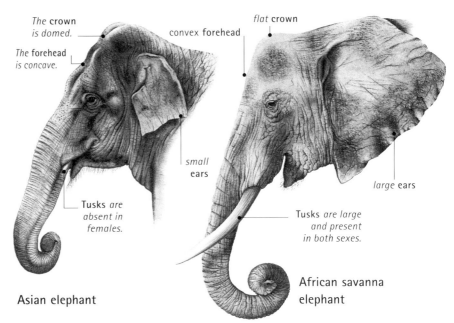

The **crown** is domed.

The **forehead** is concave.

convex **forehead**

flat **crown**

small ears

Tusks *are absent in females.*

Asian elephant

large ears

Tusks *are large and present in both sexes.*

African savanna elephant

tusks only rarely. The largest tusk ever recorded came from a male African savanna elephant; it was nearly 11.5 feet (3.5 m) long and weighed 236 pounds (107 kg). By contrast, the largest recorded tusk from a female African elephant weighed only 40 pounds (18 kg). Large tusk size in male elephants is an adaptation for combat. Males sometimes use their tusks to fight each other for the right to mate with females. Rivals can inflict severe injuries, and occasionally

▲ TRUNK AND HEAD

Head anatomy provides a number of clues for figuring out to which species an elephant belongs.

EVOLUTION

Ancient elephants

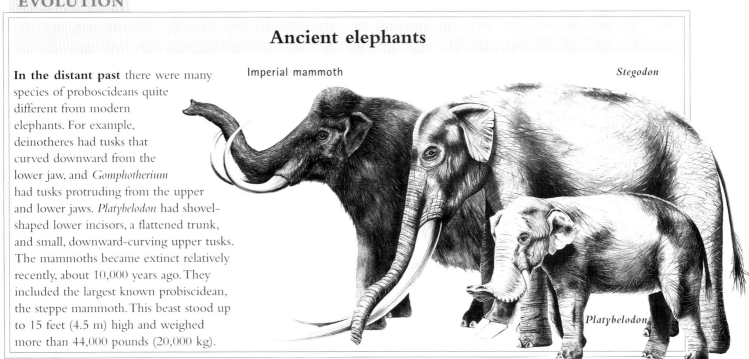

Imperial mammoth

Stegodon

Platybelodon

In the distant past there were many species of proboscideans quite different from modern elephants. For example, deinotheres had tusks that curved downward from the lower jaw, and *Gomphotherium* had tusks protruding from the upper and lower jaws. *Platybelodon* had shovel-shaped lower incisors, a flattened trunk, and small, downward-curving upper tusks. The mammoths became extinct relatively recently, about 10,000 years ago. They included the largest known probiscidean, the steppe mammoth. This beast stood up to 15 feet (4.5 m) high and weighed more than 44,000 pounds (20,000 kg).

▶ *Elephant hide is wrinkled, thick, and very tough, and is resistant to all but the most specialized parasites. Nonetheless, it can be damaged by intense sunlight. Elephants seek out shade in which to rest during the hottest part of the day.*

COMPARATIVE ANATOMY

Manatees and dugongs

A group of mammals called tethytheres lived about 50 million years ago. The tethytheres gave rise to both the Probiscidea and the Sirenia, the group that contains manatees and dugongs. Only four species of sirenians remain, though they were much more abundant and diverse in the past. All sirenians are completely aquatic; they have a large, spindle-shape body, weighing up to 2,000 pounds (900 kg). Sirenians have paddle-shape front limbs and a flat lobed or fluke-shape tail for swimming.

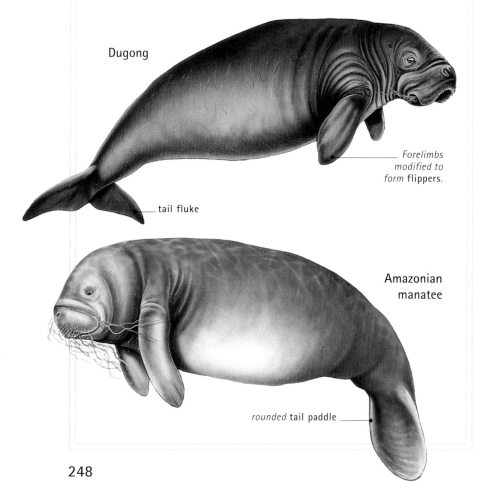

Dugong

Forelimbs modified to form flippers.

tail fluke

Amazonian manatee

rounded tail paddle

individuals may die after receiving tusk wounds in battle. Elephants also use their tusks for digging up tasty roots and for stripping bark from trees. Domesticated elephants can be trained to use their tusks to carry heavy objects such as logs.

Big ears

African savanna elephants have huge ears that can measure nearly 4 feet (1.2 m) from top to bottom. On average they are three times larger then the ears of Asian elephants. Because of their large bulky body shape, elephants sometimes have problems keeping cool in the sweltering midday heat of the tropics. Having large flapping ears helps them lose body heat, thus preventing overheating.

When elephants flap their ears, air currents cool the warm blood passing through the dense network of blood vessels lying close to the surface on the underside of the elephant's ears. They can lose enough heat through the blood vessels in their ears to lower their body temperature to comfortable levels. Elephants also use their ears to signal anger; they spread their ears wide during aggressive encounters.

Elephant skin

The skin of an elephant is wrinkled and usually gray or dull brown, although very rarely white albino elephants occur. Most of the time it is hard to see the true color of elephants because they cover themselves in mud or dust that makes them appear the same color as the local soil. Elephant skin is very thick in places, up to 1.2 inches (3 cm) on the legs, back, and trunk. Elephants sometimes suffer from sunburn. A covering of dirt helps protect the skin from the sun's rays; it may also keep biting insects at bay.

Asian elephants are born with a sparse covering of hair over their body, mostly on the top of their head and back. Most of this hair is lost after their first year. African elephants have some hair but less than their Asian relatives.

Legs and feet

Elephants sometimes have to travel long distances between feeding grounds or the water holes in which they wallow and drink. If food and water are scarce, elephants may journey across hundreds of miles to find better

COMPARATIVE ANATOMY

Hyraxes

Although they look very different, hyraxes (below) are among the closest relatives of elephants. Hyraxes resemble rabbits with short, rounded ears, but unlike rabbits hyraxes have toes with hooflike nails. Leathery footpads lubricated with glandular secretions give hyraxes amazing traction for bounding around the steep rocky slopes where they live. Hyraxes weigh up to 10 pounds (4.5 kg).

conditions. Elephants need very strong legs to move their great weight over long distances. The legs are long and powerful, and are able to support heavy loads with little effort; they direct the animals' weight through the leg bones rather than through muscles, which would quickly tire. Since they are long, the legs give the elephants an extended gait, allowing them to amble along at a steady 10 mph (16 km/h) for long periods, or at up to 25 mph (40 km/h) when charging.

Both African and Asian elephants have wide, flat feet with soft elastic soles. African elephants have five nails or "hooves" on each of their front feet and three nails on each of the hind feet. Asian elephants have five nails on each front foot and usually four nails on the back foot. Elephants have very broad feet, measuring 5 feet (1.5 m) or more in circumference. Interestingly, the shoulder height of an elephant is approximately twice the circumference of its feet, allowing trackers to estimate an elephant's size from its footprints.

▲ *The trunk has many uses besides picking up food and other objects. Water can be drawn up into the nostrils. It can then be passed into the mouth or sprayed over the body to cool the elephant down.*

Skeletal system

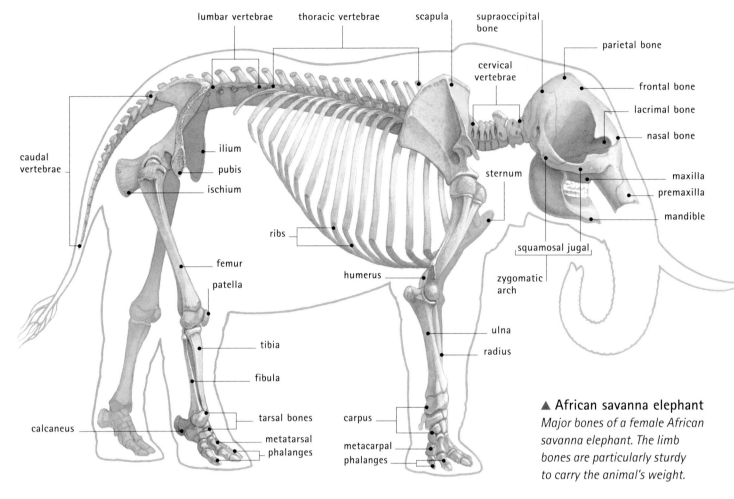

lumbar vertebrae · thoracic vertebrae · scapula · supraoccipital bone · parietal bone · cervical vertebrae · frontal bone · lacrimal bone · nasal bone · caudal vertebrae · ilium · pubis · ischium · sternum · maxilla · premaxilla · mandible · ribs · squamosal jugal · zygomatic arch · femur · humerus · patella · ulna · radius · tibia · fibula · tarsal bones · carpus · calcaneus · metatarsal phalanges · metacarpal phalanges

▲ **African savanna elephant**
Major bones of a female African savanna elephant. The limb bones are particularly sturdy to carry the animal's weight.

COMPARE the limb posture and bone structure of an elephant with that of lighter four-legged mammals such as the **RED DEER** and **WILDEBEEST**.

COMPARE the structure of an elephant's backbone with that of fast-running land mammals such as the **ZEBRA**.

Like all mammals, an elephant has a skeleton that can be divided into three major sections: the skull, the backbone and ribs, and the limb bones. The total number of bones in an elephant skeleton varies slightly between individuals and species. The skeleton as a whole is massive, accounting for between 12 and 15 percent of an elephant's body weight. The skeleton must support the elephant's great weight as efficiently as possible.

Heavy animals such as elephants have thicker, more bulky bones than lighter species. However, elephant bones are not as thick as scientists would predict on the basis of their weight, because of their unusual construction. Elephant leg bones do not have marrow-filled spaces like the bones of most other land mammals. Instead, they are filled with spongy bone tissue, which makes them stronger while keeping their weight down.

Manatees and dugongs are relatives of elephants. They have particularly heavy skeletons composed of solid bones. These animals are aquatic, and their bodies are supported by water, so the extra bone mass is not necessary for strength. The additional weight helps keep them submerged and allows them to maintain their position in the water.

The skull

An elephant's skull is very large but relatively short compared with the skulls of other mammals. With the skull bones plus the tusks, and a long muscular trunk, an elephant's head can weigh up to 660 pounds (300 kg). It takes less muscle power to hold up a short skull than a longer one, owing to reduced leverage. The main part of the skull is composed of very thick bone filled with air cavities. This makes the skull strong but relatively light.

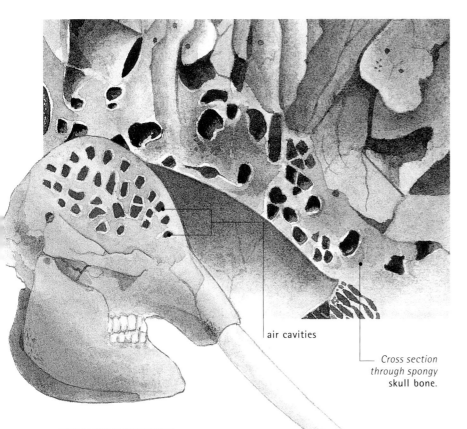

of the variation in the total number of bones between elephants comes from differences in the number of vertebrae between individual animals. The caudal section alone can contain between 18 and 33 vertebrae in the African savanna elephant and 24 to 34 vertebrae in the Asiatic elephant. The thoracic and lumbar sections of the backbone support the considerable weight of the digestive system and other internal organs, as well as the surrounding musculature. As in other types of large mammals, these sections of the backbone form a gentle arch, like a bridge. This shape provides extra strength.

Compared with backbones of smaller mammals, an elephant's backbone is relatively stiff and allows little flexibility. The stiffness maintains the backbone's arched shape and also helps reduce excessive movements of the body during locomotion, which could endanger the load-bearing bones and joints in the limbs. The thoracic vertebrae closest to the shoulder have particularly long dorsal (upper-side) processes, or projections. These are most notable in African savanna elephants; they act as anchorage points for ligaments that extend to the skull and help support the head.

air cavities

Cross section through spongy skull bone.

▲ REDUCING WEIGHT

Much of the skull contains hollow cavities formed by spongy bone. These are strong yet light compared with the denser bones of the legs. The skull is supported by ligaments and muscles attached to the thoracic vertebrae.

In the center of the large mass of skull bones is a relatively small brain chamber. The large surface area of the back of the skull provides ample attachment space for the large muscles that run to the shoulders. These muscles support the head.

The elephant's lower jaw, or mandible, is also very large. Unlike the skull it is made of solid bone. The front parts of the upper jawbone, the premaxillary bones, form sheaths for the tusks, a feature unique to elephants.

The backbone

Like all vertebrates, mammals have a backbone made up of units called vertebrae. The vertebral column can be divided into five sections: the cervical (neck), thoracic (upper back), lumbar (lower back), sacral (above the pelvis), and caudal (tail) sections. Most

Comparing skulls

Both African and Asian elephants have skulls that contain between 53 and 55 bones, depending on the individual. Although the skulls of both species have a similar structure, they have a slightly different shape. When viewed sideways, the Asiatic elephant has a concave (curving inward) forehead, while the African elephant has a convex forehead (curving outward). The skull of the African elephant is broader than that of its Asiatic cousin. Owing to their much larger tusks, African elephants have larger tusk tubes, formed by the premaxillae, on their upper jaws.

convex bone

concave bone

premaxilla

tusk

◀ Elephant skulls

Comparison of the skulls of an African savanna (left) and Asian (right) elephant. Note the differences in premaxilla and tusk size.

251

▲ *Elephants do not run. Running would place too great a strain on their limbs and joints. Instead, when speed is required they use a fast walk or jog.*

▶ FOOT
African savanna elephant
Surprisingly, elephants walk on their toes. An elastic pad supports and cushions the rear of the foot against impact.

tarsal bones

metatarsal

Phalanges. *Each foot bears five toes, though the side ones are often much reduced.*

nails

tibia

calcaneus

The pad is made of fatty, elastic tissues.

The sole of the foot is covered by a thick layer of skin. Skin is replaced as soon as it is sloughed.

The limbs

An elephant's leg bones are massive, relatively thicker than the leg bones of smaller mammals. As with any column-shape object, the bone of a leg is better at withstanding the compressive forces that run along the length of the bone than the bending forces operating across it. Since elephant bones have straight shafts and are aligned vertically, the weight of the animal passes through the length of the bones. To prevent shearing (tearing) forces at bone junctions, all the joint surfaces are in line with the bone shafts.

Elephant limb bones have a number of features that help the animals maintain their vertical alignment. As in all mammals, an elephant's principal limb bones are the humerus (upper forelimb), the radius and ulna (lower forelimb), the femur (thighbone), and the fibula and tibia (lower hindlimb). Unlike humans, the elephant has a scapula (shoulder blade) positioned directly above the humerus and not held in place by a clavicle or collarbone—this is absent in elephants. The heads of the humerus and femur point nearly straight upward, so they transfer weight directly down the length of the leg bone. Compare this with the head of the femur in the human skeleton, which sticks out sideways and connects with the hipbones (pelvis).

Wrists and ankles

In the skeletons of mammals that specialize in running, such as horses and cheetahs, the wrist bones, or carpals, and ankle bones (the tarsals) may be as long as, or longer than, the other limb bones. This allows them to provide enormous leverage. By contrast, the carpal and tarsal bones of elephants are small and blocklike. Elephants never actually run, although they can move surprisingly fast when necessary, adopting a kind of bustling jog. The heavy jolting that occurs in running would cause injury to the animal's joints, bones, and muscles.

The feet

An elephant has five toes on each foot, although the outer pair of toes may be very small in some individuals. Unlike humans, who walk with their heels on the ground, an elephant stands on what corresponds to the ball of the foot. This kind of stance is called digitigrade (from Latin, meaning "finger walking"). Most of an elephant's weight rests on a broad pad of elastic tissue behind the toes. This acts as a shock absorber and prevents the skeleton from jolting too much when the elephant walks. It also allows elephants to move surprisingly quietly despite their size.

▼ Elephants are social animals. Several anatomical adaptations, from the structure and location of the hyoids to a resonating pouch in the throat, allow them to produce a range of sounds for communication.

The hyoid bones

Like other mammals, elephants have a group of bones in their throat known as hyoids. Elephants have five of these small, delicate bones. The hyoids are involved in several functions such as feeding and sound production, including the infrasonic rumblings that elephants use to communicate over long distances.

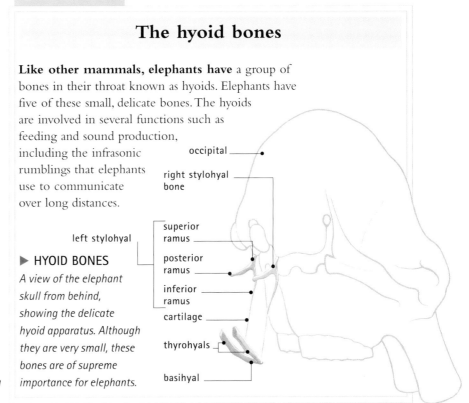

left stylohyal

▶ HYOID BONES

A view of the elephant skull from behind, showing the delicate hyoid apparatus. Although they are very small, these bones are of supreme importance for elephants.

occipital

right stylohyal bone

superior ramus

posterior ramus

inferior ramus

cartilage

thyrohyals

basihyal

Muscular system

COMPARE the relatively small jaw muscles of a herbivore like an elephant with those of a *HUMAN*.

COMPARE the large ear muscles of an elephant with those of a *WOLF*.

Elephants, like other vertebrates, have three main types of muscles: skeletal or striated muscle, smooth muscle, and cardiac muscle. Skeletal muscles are the large muscles attached to the skeleton; they are mostly used for locomotion and other body movements and to maintain posture. When viewed under a microscope, skeletal muscle fibers are seen to have distinctive dark and light bands, or striations. Smooth muscles are present in many internal organs, such as the esophagus, stomach, and intestines, and in blood vessels. These muscles are composed of relatively short cells and are responsible for moving food down the throat during swallowing and moving food through the intestines. Cardiac muscle occurs only in the walls of the heart. It is responsible for pumping blood around the circulatory system.

▼ **African savanna elephant**
Powerful muscles in the limbs and neck are vital for locomotion and for moving the head.

▶ **HEAD AND NECK**
An elephant uses the auriculo-occipitalis to flap its ears.

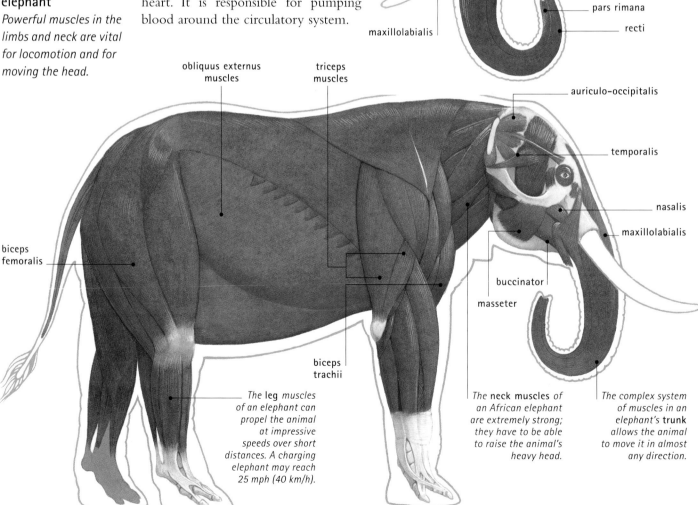

The **leg** muscles of an elephant can propel the animal at impressive speeds over short distances. A charging elephant may reach 25 mph (40 km/h).

The **neck** muscles of an African elephant are extremely strong; they have to be able to raise the animal's heavy head.

The complex system of muscles in an elephant's **trunk** allows the animal to move it in almost any direction.

A lack of rotation

The muscular systems of African and Asian elephants are virtually identical, and they are structurally similar to those of most other mammals. For example, there are 41 muscles in an elephant's foreleg; most of these occur in a human arm, although their relative proportions may differ. However, unlike humans and most other mammals, elephants are unable to rotate their forelimbs. The inability to rotate the forelimbs is one of the features that define the order Proboscidea and is probably an adaptation for reducing the risk of injury to heavy weight-bearing leg joints. Two of the forelimb muscles used by mammals to rotate the foot, the pronator quadratus and the pronator radii teres, are small or absent in elephants. The other forelimb muscles—such as the biceps and brachialis, which flex the front leg; and the triceps muscles, which extend the front leg—are also found in the human arm. Many skeletal muscles, such as the triceps and biceps, act in antagonistic pairs; they work against each other in opposite directions.

Size and shape

Since elephants have a different posture from humans, the relative sizes and shapes of their muscles are different from those of their human equivalents. For example, the foreleg extensor muscles, or triceps, which help maintain the elephant's upright posture during walking, are relatively powerful. Elephants also have particularly large neck muscles to maintain their heavy head in an upright position. There are 48 muscles in the elephant's hind legs, the largest being the principle hamstring muscle, the biceps femoris. The hamstring muscles flex the back legs and are used to provide much of the forward thrust of a walking elephant.

Excluding the trunk, elephants have 140 muscles in the head. These muscles determine head position and control the movement of the jaws, lips, and eyes. Unlike humans, elephants are able to use well-developed muscles (the auriculo-occipitalis) for flapping their large ears. There are only 16 main muscles around the body wall, including some of the largest in the elephant's body. The body-wall muscles maintain posture and help hold in the intestines and other internal organs.

Muscles of the trunk

An elephant's trunk contains no bone or cartilage and is made up mostly of muscle. Elephants can move their trunk in almost any direction and hold it in a variety of positions. The trunk is strong enough to pick up a large log yet has enough dexterity to manipulate small stones. To be able to accomplish such contrasting tasks, the trunk needs a complex muscular structure. French anatomist Georges Cuvier (1769–1832) estimated that there were about 40,000 muscles in an elephant's trunk. Modern anatomists, however, believe that many of these are subunits, or fascicles, from seven major muscle groups.

▼ TRUNK
African savanna elephant
This lengthwise cross section shows the main groups of muscles of the trunk.

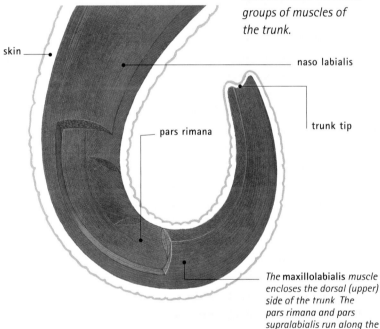

skin

naso labialis

trunk tip

pars rimana

The maxillolabialis *muscle encloses the dorsal (upper) side of the trunk. The pars rimana and pars supralabialis run along the length of the underside.*

► TRUNK CROSS SECTION African savanna elephant

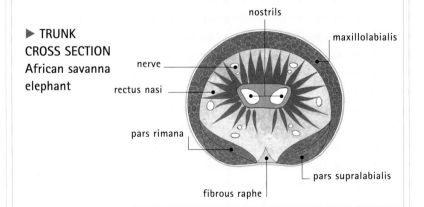

nostrils

maxillolabialis

nerve

rectus nasi

pars rimana

pars supralabialis

fibrous raphe

255

Nervous system

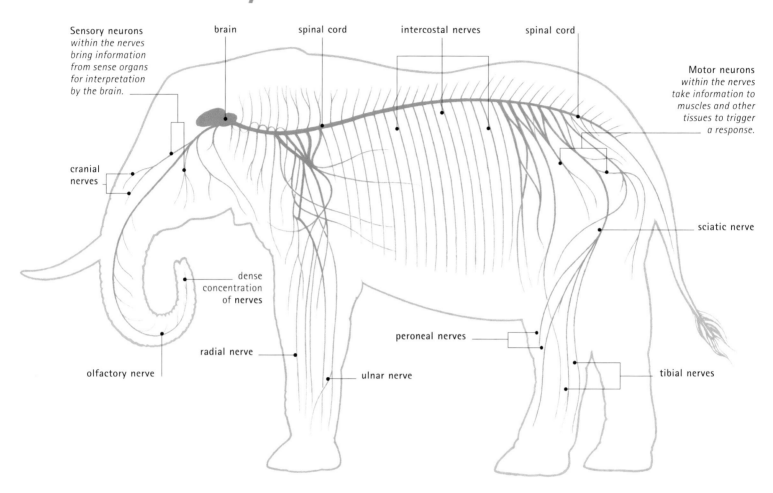

Sensory neurons *within the nerves bring information from sense organs for interpretation by the brain.*

brain

spinal cord

intercostal nerves

spinal cord

Motor neurons *within the nerves take information to muscles and other tissues to trigger a response.*

cranial nerves

sciatic nerve

dense concentration of nerves

radial nerve

peroneal nerves

olfactory nerve

ulnar nerve

tibial nerves

▲ **African savanna elephant**
Nerves near the tip of the trunk connect with the olfactory nerve. This arrangement gives the trunk its extreme sensitivity.

CONNECTIONS

COMPARE the shape and relative size of an elephant's cerebrum with that of a **HUMAN**.

COMPARE the sound-based communication of an elephant with that of a **DOLPHIN**.

Like all mammalian nervous systems, an elephant's nervous system can be divided into two main parts: the central nervous system (CNS) and the peripheral nervous system (PNS). The CNS includes the brain and the spinal cord. The PNS includes all the other nerves that lead to and from the CNS. Both the brain and the spinal cord have a hollow cavity at the center, though the elephant's spinal cord cavity is notably reduced compared with that of some other mammals. The spinal cord of all vertebrates passes through the middle of the vertebral column.

Types of cells
The nervous system is made up of specialized cells called neurons, which are able to transmit electrical signals. There are three main types of neurons. Sensory neurons conduct nervous impulses from sense organs such as the eyes, ears, and skin to the central nervous system.

Motor neurons carry signals to the muscles and initiate muscle contraction. Finally, interneurons connect the sensory and motor neurons. Nerves are bundles of neurons along with small blood vessels that bring nutrients and oxygen to the neurons. Like all other vertebrates, elephants have pairs of large nerves leaving each side of the spinal cord at regular intervals. These nerves divide into many branches, leading to specific parts of the body. For example, the nerves leaving the thoracic region of the spine carry sensory and motor neurons to and from organs such as the heart, lungs, stomach, liver, pancreas, and kidneys, as well as many of the muscles of the neck and chest region.

The brain
Elephants have the largest brain of any land mammal, although it is not especially large relative to the size of the body. An

African savanna elephant's brain weighs up to 14 pounds (6.5 kg). It is slightly larger than that of the Asian elephant, which weighs up to 12 pounds (5.5 kg). An elephant's brain can be divided into five distinct parts. The largest is the cerebrum, which in turn is divided into two halves called cerebral hemispheres. In relatively intelligent animals, such as elephants, the cerebrum is the largest part of the brain.

The outer layer of the cerebrum (the cerebral cortex) is highly folded in elephants, resulting in an increased surface area and many more neurons. The cerebral cortex gives the brain its characteristic appearance. It is made up of gray matter comprising billions of neurons. The cerebral cortex of the human brain has 100 billion neurons, and the figure for elephants is probably similar. The cerebral cortex determines how sensory information is interpreted, controls voluntary movements, and is the seat of consciousness, memory, and learning. Another notable feature of an elephant's brain is the large cerebellum, which is located behind the cerebrum. In mammals, the cerebellum is important for coordinating movement and balance.

The sense organs

Elephants gather information from the world around them using their senses of sight, hearing, smell, touch, and taste. For elephants, the sense of hearing is particularly important, and they are able to communicate over distances of several miles using low-frequency

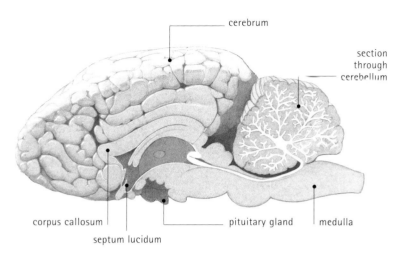

cerebrum

section through cerebellum

corpus callosum

septum lucidum

pituitary gland

medulla

sound, or infrasound. Infrasound is significant for African forest elephants, because unlike higher-frequency sounds, infrasound can travel relatively long distances through forest vegetation. Elephants are also capable of detecting ground vibrations through their feet. This allows them to keep in contact with herds up to 10 miles (16 km) away.

Elephants have a very keen sense of smell and can detect chemical signals called pheromones left by other elephants. Elephants are also able to smell approaching predators. When elephants catch the scent of an enemy, they raise their trunk to sniff the air. If the danger is close they form a protective circle around their young. Elephants are particularly sensitive to touch, especially on the tip of their trunk, the skin of which has a very high density of touch receptors.

▲ BRAIN
African savanna elephant
The brain is very heavy, weighing up to 12 pounds (5.4 kg) in males and 9.5 pounds (4.3 kg) in females. Also, the active surface area of the brain is large, suggesting intelligence.

▼ SENSITIVE TRUNK
African savanna elephant
The tip of the trunk is highly sensitive. Fine hairs are able to feel even tiny objects.

Tool-using tuskers

The ability to make and use tools is considered to be one of the most important signs of intelligence in animals. Many animals use tools, but elephants are unusual because they also modify objects to make them more effective as tools. A study of Asian elephants in India showed that many captive and wild elephants used switches made of branches to keep flies away. Some of these elephants modified branches by stripping some of the leaves and shortening the stem to make them into more effective fly switches. Using their trunk, elephants sometimes pick up stones and other material to throw at enemies. In one case, a female African savanna elephant was observed throwing mud and sticks at a white rhinoceros that was bothering her.

Fine sensory **hairs** *around the tip of the* **trunk** *connect to the nervous system, passing information about the environment to the brain.*

Circulatory and respiratory systems

Elephants have the same basic plan for their circulatory and respiratory systems as other mammals, but because of their large size these systems include some unusual features.

The circulatory system

In common with other mammals, elephants have a four-chamber heart, consisting of two ventricles and two atria. Arteries carry blood away from the heart, while veins carry blood toward the heart. Deoxygenated blood from the elephant's body passes into the right atrium via a pair of thick veins, the vena cavae. Contractions of the right atrium force blood through a one-way valve into a larger, more powerful, chamber, the right ventricle.

The right ventricle then squeezes the blood via the pulmonary artery through the small blood vessels, or capillaries, of the lungs. There, red blood cells release waste carbon dioxide and pick up oxygen. Oxygenated blood returning from the lungs passes into the left

atrium, which pushes the blood through another one-way valve into the most powerful chamber of the heart, the left ventricle. The left ventricle needs to be powerful to pump the oxygenated blood through the main artery, the aorta, and onward around the rest of the elephant's body.

A peculiar feature of the elephant's heart is that the two ventricles are separated near the top of the heart, giving it the appearance of a "double heart." The only other mammals to share this characteristic are the sirenians, elephants' closest relatives. The heart of a large bull African elephant can weigh up to 60 pounds (27 kg). This might seem very large, but it is not excessively heavy compared with an elephant's body size.

Cells and vessels

An African savanna elephant can have up to 200 gallons (750 l) of blood, accounting for 10 percent of its body weight. Like almost all

▼ CIRCULATORY SYSTEM African savanna elephant
Despite their large size, elephants display similarities with other mammals in their respiratory and circulatory systems.

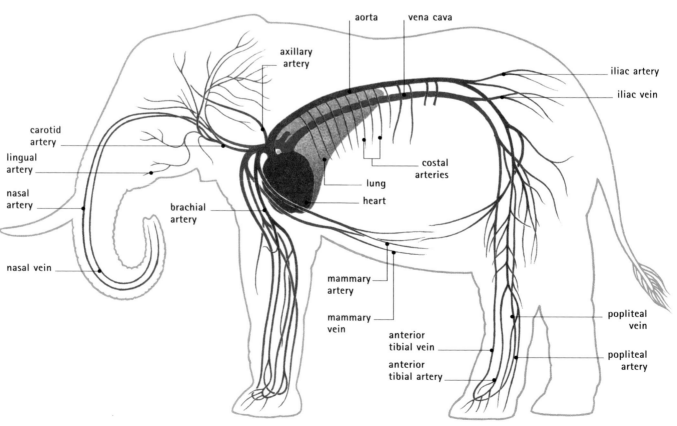

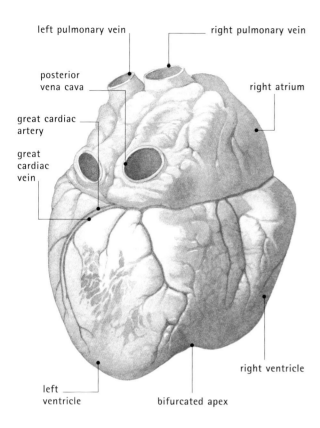

left pulmonary vein

right pulmonary vein

posterior
vena cava

right atrium

great cardiac
artery

great
cardiac
vein

right ventricle

left
ventricle

bifurcated apex

▲ *Just as in smaller mammals, the heart accounts for about 0.5 percent of the body weight. However, in elephants, unlike most other mammals, the two ventricles are divided near the top of the heart.*

vertebrates, elephants have red blood cells containing the red pigment hemoglobin. Hemoglobin binds to oxygen so it can be transported around the body. Elephant hemoglobin has a higher affinity for oxygen than that of humans.

Elephant arteries are large and heavy, and are supported by ridges of connective tissue or muscles. Elephants need thick-walled arteries so they can withstand a high blood pressure, nearly 50 percent higher than that of humans. Elephant veins are also thick-walled; this keeps these wide blood vessels from collapsing. Some blood vessels in an elephant reach up to 11 feet (3.5 m) long.

Lymphatic system

The elephant circulatory system also includes the lymphatic system. The lymphatic system is a network of vessels transporting fluid and plasma proteins that leak from blood capillaries into surrounding body tissues. The lymphatic system takes these materials back to the main circulatory system. Lymph nodes occur in places where small lymphatic vessels come together. The lymph nodes are important sites for the production of white blood cells, which combat invading microorganisms and other objects in the body.

The respiratory system

Elephants can breathe through their mouth or their trunk. This ability enables them to breathe when their mouth is full or when they are using their trunk to suck up water or dust. Air passes down the trachea (windpipe) and into two lungs via the bronchial tubes. Elephants inhale around 80 gallons (310 l) of air each minute.

Elephants and most other mammals have a space between the lungs and the chest wall called the pleural cavity. Raising the ribs and lowering a membrane called the diaphragm will increase the volume of the pleural cavity, creating a negative pressure that causes the lungs to inflate. An elephant's pleural cavity is filled with a stretchy network of collagen fibers that connects the lungs to the chest wall and the diaphragm. Elephants rely more on the diaphragm and less on the muscles raising the rib cage than other mammals. Elephant lungs have a network of thick elastic tissues that prevents the alveoli (air sacs) from being squashed by the mass of surrounding tissues. These unusual features help the animals overcome being heavy and allow them to snorkel (using their trunk for breathing) in deep water.

IN FOCUS

Heart rate and energy efficiency

The heart rate of an elephant is about 30 beats per minute, which is considerably slower than that of smaller mammals. For example, an average human has a resting heart rate of 70 beats per minute, and a mouse has a heart rate of 500 beats per minute. An elephant can afford to have a comparatively slow heart rate, and hence a relatively slow rate of oxygen delivery to the body, because pound for pound it needs less energy than smaller mammals. This is because the large bodies of elephants conserve heat well, so they do not need to generate as much heat as small mammals do. Elephants are also able to tolerate a fairly wide range of body temperatures. In addition, elephants use relatively little energy in moving around because of the efficiency of their skeleton and musculature.

Digestive and excretory systems

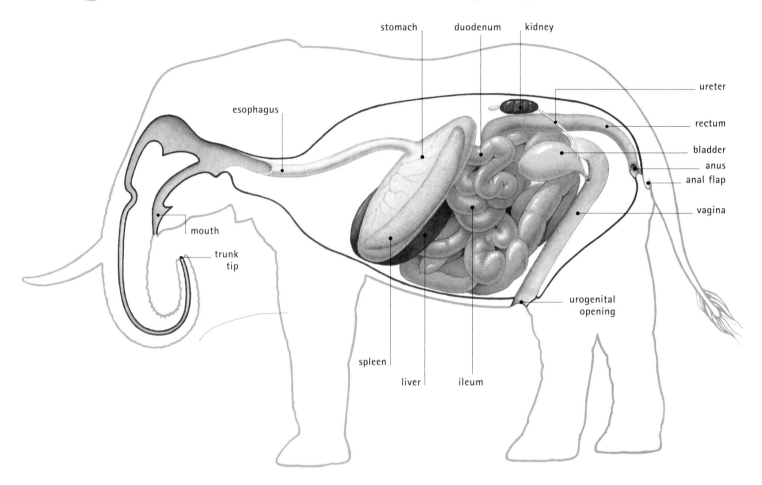

stomach   duodenum   kidney

esophagus

ureter

rectum

bladder

anus

anal flap

vagina

mouth

trunk tip

urogenital opening

spleen

liver   ileum

▲ **Female African savanna elephant**
The very long intestine is necessary for the digestion and absorption of nutrients from woody vegetation.

CONNECTIONS

COMPARE the structure of an elephant's stomach with that of ruminants such as a *WILDEBEEST* or a *RED DEER*, which have a complex multichamber stomach for the breakdown of tough cellulose.

Elephants eat a range of vegetation that includes grass, leaves, fruit, and woody material such as twigs, roots, and even bark. Asian elephants sometimes raid crops such as sugarcane and bananas. Considering that Asian elephants can eat about 330 pounds (150 kg) of food each day, it is not surprising that crop-raiding elephants are unpopular with farmers.

The advantage of size
In dry habitats, much of an elephant's diet is made up of tough fibrous plant material, particularly twigs. Even though this food is not very nutritious and is difficult to digest, elephants are able to live on it because of their very large size and relatively low energy requirements. Being large, an elephant can have the long and voluminous digestive system needed to digest and absorb woody vegetation. An elephant's intestines can measure up to 115 feet (35 m) long.

The digestion process starts in the mouth, where vegetation is chewed and mixed with saliva. Elephants chew with great molars that can weigh up to 11 pounds (5 kg) each. The surfaces of these teeth bear enamel ridges that help grind vegetation. Unlike the teeth of

IN FOCUS

A gargantuan thirst

Elephants prefer to drink every day, although they can go for several days without drinking water if necessary. A large male will drink about 58 gallons (220 l) of water in a day—and can take up to 26 gallons (100 l) in one go. Elephants obtain water from rivers and pools, or from groundwater, which they get at by digging pits with their tusks and trunk.

other mammals, which are replaced with new teeth erupting from below, new elephant teeth come from behind the old ones as if on a conveyor belt. An elephant can have six sets of teeth throughout its 50- to 70-year lifetime.

Elephants have well-developed salivary glands and also mucus glands in the esophagus that moisten dry vegetation, allowing it to move easily down to the stomach. The elephant digestive system is simple compared with those of other plant-eating mammals. Elephants have a vertical, cylinder-shape stomach that acts as a food-storage chamber.

Micro-partners

Cellulose is a tough chemical found in the walls of plant cells. Some plant-eating animals, such as ruminants (antelopes, deer, and their relatives) have a multichamber stomach where the cellulose is broken down by micro-organisms. Elephants also need the help of symbiotic microorganisms to digest cellulose. Rather than taking place in the stomach, this process occurs in a chamber called the cecum. The cecum lies at the junction of the small and large intestines. Nutrients released by cellulose breakdown are absorbed directly through the cecum wall, which contains many blood vessels to transport the nutrients around the body.

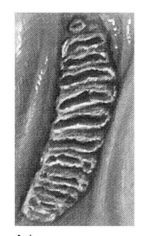

African savanna   Asian

◀ *A molar tooth of an African savanna elephant and an Asian elephant. Key differences allow the teeth to be told apart. The African elephant's molars are lozengelike, while the Asian elephant's are more rectangular. There are also differences in the pattern of enamel ridges on the teeth.*

Elephant dung

The indigestible remainder of food passes from the cecum into the large intestine, from where water is reabsorbed. Then food passes to the rectum. There, feces are stored until they are ejected as droppings. Elephants can produce 240 pounds (110 kg) of dung per day.

Except for their large size, the other organs involved in digestion and excretion in elephants, such as the liver, kidneys, and pancreas, are typically mammalian. Elephants do not have a gallbladder, but instead have an enlarged hepatic duct that pipes bile from the liver to the intestines.

IN FOCUS

A keystone species

Partly because of their size, elephants play an important role in the ecosystems in which they live. Elephants effectively shape the plant communities around them by selectively browsing certain plants and by killing trees. They do this by pushing trees over to get at leaves high in the canopy, or by stripping bark. Elephants also act as seed and nutrient dispersers. Because of elephants' relatively inefficient digestive system, their dung is full of nutrients and often contains seeds. Elephant dung is an important resource for insects such as flies and dung beetles; the young insects feed on the droppings.

▶ *Nutrient-rich elephant dung provides a food source for a wide range of animals, plants, and fungi.*

Reproductive system

Both African and Asian elephants are slow breeders. Female, or cow, Asian elephants do not become sexually mature until eight years old, and maturity does not occur until the age of 11 in female African savanna elephants. In all elephant species, males reach sexual maturity even later. They do not usually have the opportunity to mate until they are much older when they have a higher social position. Female elephants usually give birth to a single calf every five years or so, depending on conditions.

The estrus cycle

Like all placental mammals, female elephants produce and release eggs at regular intervals in response to changing hormone concentrations in the blood. When the eggs are released from the ovaries, they pass down the fallopian tubes into the uterus ready to be fertilized. At this time females become much more sexually receptive. The time between periods of sexual

▶ **FEMALE REPRODUCTIVE ORGANS African savanna elephant**
The vaginal tract opens between the female's hind legs.

▼ **MALE REPRODUCTIVE ORGANS African savanna elephant**
The elephant's penis is S-shaped and is the largest of any land mammal.

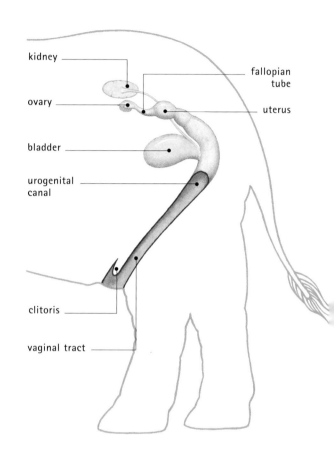

kidney

ovary

bladder

urogenital canal

fallopian tube

uterus

clitoris

vaginal tract

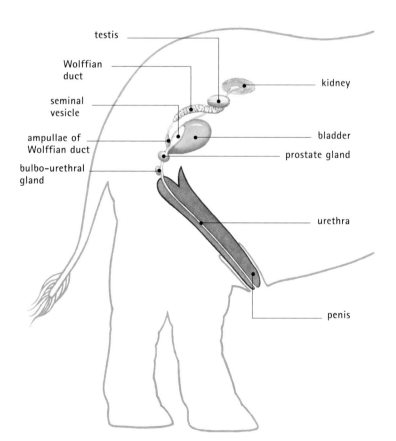

testis

Wolffian duct

seminal vesicle

ampullae of Wolffian duct

bulbo-urethral gland

kidney

bladder

prostate gland

urethra

penis

Meeting mates

Elephants live in family groups made up of several related adult female elephants and their young. Each family group is led by a dominant female. Adult bulls spend much of their time away from these family groups, and usually join a group only to mate. Adult females in breeding condition, or estrus, make low-frequency calls inaudible to humans to attract males from up to several miles away. Sometimes the older bull elephants become very aggressive and actively seek out females in estrus. This period, which can last from a few days to a few months in more mature bulls, is called musth.

receptivity is called the estrus cycle. In Asian elephants, this period is around 22 days, and in African savanna elephants it is about 60 days.

Female reproductive organs

Apart from their large size, the reproductive organs of female elephants are typical for a mammal. The embryo develops in the uterus, which is located under the pelvis. The time from egg fertilization to birth (the gestation period) is around 22 months. At birth, the baby elephant, which weighs around 265 pounds (120 kg), passes through the vaginal opening. This lies between the elephant's hind legs; the vaginal openings of most other large quadrupedal (four-legged) mammals, such as cattle and deer, is under the tail.

Females have a pair of milk-producing mammary glands. Unlike the mammary glands of other quadrupedal mammals, those of elephants are located between the front legs. Mother and calf can maintain trunk contact during feeding, helping strengthen their bond.

Male reproductive anatomy

Like other male mammals elephants have two testes and a penis. During mating with a female, sperm passes from the testes along the genital ducts to the penis, from where it passes into the female's vagina. The testes of male elephants are located inside the elephant's

CLOSE-UP

Male musth

At around the age of 20, bull elephants experience a phase called musth, which lasts two or three months every year. Bull elephants in musth produce large amounts of fluid from the temporal glands, which are located behind each eye. Cow elephants in estrus usually choose a large bull in musth and ignore the smaller males. Males in musth are chosen over other bulls, even those larger than themselves. Bull African savanna elephants can be particularly aggressive during musth periods, and occasionally they inflict fatal tusk wounds on their opponents.

body, close to the kidneys. This is unusual in mammals, though the elephant's relatives, the sirenians—dugongs and manatees—also have internal testes.

A male elephant's penis is controlled by powerful muscles and can, when erect, reach more than 3 feet (1 m) long in mature males. When not erect, the penis is retracted within the elephant's body.

ADRIAN SEYMOUR

FURTHER READING AND RESEARCH

Hare, T. and M. Lambert. 1997. *The Encyclopedia of Mammals.* Marshall Cavendish: New York.
MacDonald, David. 2006. *The Encyclopedia of Mammals.* Facts on File: Tarrytown, NY.

▼ *A newborn elephant calf may consume 24 pints (11.4 l) of its mother's milk every day.*

CLOSE-UP

Elephant pheromones

Like many mammals, elephants communicate their sexual condition to potential mates and competitors using pheromones. Pheromones are volatile chemicals in body secretions such as urine or sweat. Elephants detect pheromones using their powerful sense of smell. Bull elephants in musth pass large amounts of urine with strong pheromones. While these pheromones may serve to attract dominant females in estrus, subordinate females or females with calves will back away from these secretions or even display defensive behavior.

Endocrine and exocrine systems

Animals use nerves to send messages around the body within a fraction of a second, to regulate functions such as pain sensation or muscular activity. However, many life processes take place over much longer time scales. Many of these processes—both in animals and in plants—are controlled by hormones. Hormones are chemical signals that initiate some action in the body. They control the day-to-day functions of the body, such as digestion, reactions to stress, and regulation of sugar levels in the blood. Hormones also regulate longer-term physical changes, such as growth and development.

Hormones are produced by a network of glands called the endocrine system. Hormones act inside the body. Not all secretions from glands are hormones, though. A separate network called the exocrine system releases chemicals to the outside of the body and onto the surfaces of cavities inside it. Sweat and salivary glands, for example, are important parts of the exocrine system.

Nerves, glands, and the endocrine system

All organisms that have a nervous system also have some type of endocrine system. When body conditions need fine-tuning, the nervous system signals the glands or glandlike tissues of the endocrine system. The glands then release the relevant hormones. Most endocrine glands in vertebrates are tightly packed cells that form thin layers called epithelial tissues. Glands are generally small and are well supplied with blood vessels; they rely on blood flow to carry their chemical signals around the body. The glands of fish, amphibians, and reptiles tend to be diffuse, spread throughout a tissue. In mammals and birds, most glands consist of a single clump of secretory cells.

Most endocrine glands have no ducts but are surrounded by other tissues. These ductless glands secrete hormones directly into the spaces between cells, from where they drift into the bloodstream. By contrast, most exocrine glands have ducts that channel their secretions away.

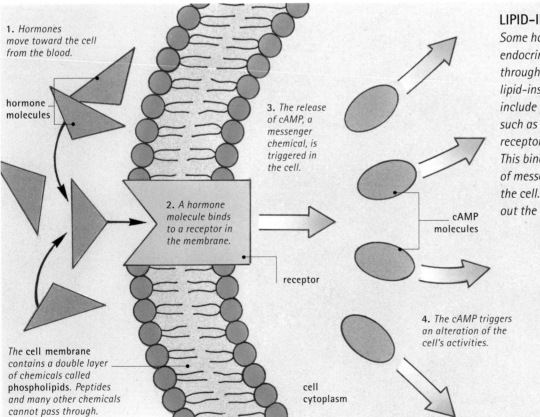

1. Hormones move toward the cell from the blood.

hormone molecules

3. The release of cAMP, a messenger chemical, is triggered in the cell.

2. A hormone molecule binds to a receptor in the membrane.

receptor

The cell membrane contains a double layer of chemicals called phospholipids. Peptides and many other chemicals cannot pass through.

cell cytoplasm

cAMP molecules

4. The cAMP triggers an alteration of the cell's activities.

LIPID-INSOLUBLE HORMONES
Some hormones released by endocrine glands cannot pass through a cell wall. They are called lipid-insoluble hormones, and include protein-based hormones such as insulin. They bind with receptors on the cell membrane. This binding results in the release of messenger chemicals inside the cell. They make the cell carry out the hormonal instructions.

Vertebrate hormones transmit their messages in three ways. In autocrine signaling, the hormone acts on the cell that secretes it. In paracrine, or local, signaling the hormone acts on nearby cells. In endocrine, or long-distance, signaling, the hormone circulates in the blood and acts on distant cells.

How hormones work

Each hormone can act only on its target cells. They have receptors that can bind only to that hormone. Nontarget cells are also exposed to the hormone. However, they do not have the correct receptor, so they cannot respond to the hormone. Hormones are either lipid- (fat-) soluble or lipid-insoluble. Lipid-soluble hormones can penetrate cell membranes and bind to receptors inside the cell. Lipid-soluble hormones often have long-term effects on the body, controlling growth and development. Lipid-insoluble hormones are not able to enter cells. Instead, they bind to receptors on the cell's outer membrane.

When a hormone binds to its receptor, the cell is induced to perform a specific function. Some receptors do this directly by moving into the cell with the hormone molecule attached. Others activate new chemicals that carry the message into the cell. Either way, a cascade of chemicals is released that carries out the hormone's instructions.

One of the most important messenger chemicals inside a cell is called cyclic adenosine monophosphate, or cAMP. When a hormone such as glucagon or calcitonin binds to a receptor, cAMP is produced in the cell's membrane. The cAMP moves into the cell, where it triggers proteins called enzymes. They drive chemical reactions in the cell, and so control the way the cell behaves.

▲ In contrast to the endocrine system, the exocrine system produces secretions that may act externally. This strawberry poison arrow frog, for example, has glands that secrete a powerful and deadly poison onto the skin's surface. This wards off predators. The poison is produced by exocrine skin glands.

Types of hormones

The hormones produced by mammals fall into four main categories. Amines, such as thyroid hormones and catecholamines, are derived from amino acids. Eicosanoids, such as prostaglandins, are produced from chemicals called fatty acids. These hormones are stored in cell membranes and are released as soon as they are needed.

Most hormones belong to the other two groups: steroids and protein-based hormones. Steroids are lipids (fats) derived from cholesterol. They include estrogen and testosterone, which are made and secreted by glands in the sex organs. These hormones travel in the blood, and are not stored by cells. Steroid hormones are able to pass easily through the outer membranes of cells. This ability allows them to reach their specific receptors inside.

Protein-based hormones range from small peptide hormones composed of as few as three tiny amino acid molecules to giant, complex protein molecules. Insulin and the pituitary hormones are protein-based. Unable to enter cells directly, peptide hormones are stored as granules in glands and are released into the bloodstream in bursts.

SYSTEM HIGHLIGHTS

GROWTH AND DEVELOPMENT Growth from infancy through adulthood is controlled by hormones such as somatotropin. Hormonal activity drives even more dramatic changes in animals such as frogs and insects. *See pages 266–267.*

HOMEOSTASIS The body's internal environment—for example, temperature, salt concentration, and water level—is under strict hormonal control. *See page 268.*

REPRODUCTION Hormones control the development of the sex organs; they also control features such as the release of eggs from the ovaries. *See page 269.*

VERTEBRATE ENDOCRINE SYSTEMS Regulated by the hypothalamus and pituitary gland, vertebrate endocrine systems drive a suite of complex interactions. *See pages 270–273.*

EXOCRINE SYSTEM Exocrine glands release their important secretions onto the body surface or into body cavities. *See pages 274–275.*

Growth and development

The hypothalamus secretes chemicals called neurohormones. They travel down the portal vein to the anterior pituitary, causing it to release particular hormones into the bloodstream.

In all vertebrates, invertebrates, plants, and even fungi, hormones control crucial body functions, such as growth, development, homeostasis (the regulation of an organism's systems), and reproduction.

Growth hormones

Every organism needs to grow. The regulation of growth by hormones occurs in even the smallest multicellular organisms. Vertebrate growth is controlled by the hypophysis, a master gland that occurs in the brain of most

animals. In humans, this corresponds to the pituitary gland, a bean-sized gland. It lies in the brain just below another gland called the hypothalamus. Most animals have a structure called a hypothalamus or a similar gland in the brain that secretes neurosecretory hormones. These work directly on the hypophysis. Some neurosecretory hormones trigger hormone release from the hypophysis; others prevent hormone release.

Hypophysis structure is remarkably similar in most vertebrates. In most tetrapods (four-limbed vertebrates) the hypophysis consists of two sections called the pars intermedia and the pars distalis. These sections correspond to the posterior and anterior pituitary glands in humans. A growth hormone called somatotropin is secreted by the pars distalis on orders from the hypothalamus. In humans, somatotropin is synthesized and released from the anterior pituitary gland.

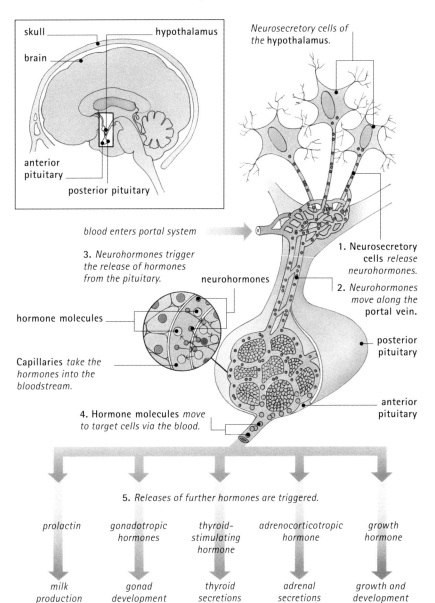

skull

brain

hypothalamus

anterior pituitary

posterior pituitary

Neurosecretory cells of the hypothalamus.

blood enters portal system

3. Neurohormones trigger the release of hormones from the pituitary.

neurohormones

hormone molecules

Capillaries take the hormones into the bloodstream.

4. Hormone molecules move to target cells via the blood.

1. Neurosecretory cells *release neurohormones.*

2. *Neurohormones move along the portal vein.*

posterior pituitary

anterior pituitary

5. Releases of further hormones are triggered.

prolactin	gonadotropic hormones	thyroid-stimulating hormone	adrenocorticotropic hormone	growth hormone
milk production	gonad development	thyroid secretions	adrenal secretions	growth and development

IN FOCUS

Growth in plants

Plants have growth hormones, too. Auxin is a hormone produced by cells in the growing regions of a plant. Auxin is plentiful at the tips of shoots, where it controls their elongation and upward growth. It is also present in the roots, triggering them to grow down into the soil.

Other plant hormones include gibberellins, which occur on leaf tips and work with auxin to elongate cells during plant growth. In some plants, gibberellins also stimulate the production of enzymes used in metabolism. These enzymes include cytokinins, which regulate cell division and, with auxin, cause cell differentiation (the formation of distinct types of cells) to take place. Other plant hormones include ethylene, which stimulates the ripening of fruit and the germination of seedlings; and abscisic acid, which regulates the falling of leaves in deciduous trees.

Keeping growth in check

Growth cannot occur unhindered. The hormone that induces growth must itself be controlled. In humans, a hormone called somatostatin inhibits growth. Most animals have growth-inhibiting hormones. Bony fish have a gland called a urophysis gland, which is located near the base of the tail. The urophysis produces two hormones (urotensins I and II) that perform the same function as the growth-inhibiting somatostatin in humans.

The hypothalamus produces a thyroid-releasing hormone (TRH); this stimulates the hypophysis to produce thyroid-stimulating hormone (TSH). This in turn acts on the thyroid gland, which is located in the neck in humans. TSH activates the production and secretion into the blood of the thyroid hormones, T3 and T4, both of which are crucial for growth and development. Similar thyroid hormones control growth in cephalopods such as octopuses.

Making adults

Thyroid hormones play a vital role in the metamorphosis (change) of frogs from tadpoles into adults. Adult frogs are insect-eating, tailless, air-breathing, and four-legged; but tadpoles are vegetarian, long-tailed, gilled, and legless. A thyroid hormone, thyroxine, controls metamorphosis in frogs. When it is young, a tadpole's body contains very little thyroxine. When the tadpole is ready to metamorphose, a signal from the brain causes the thyroid gland to begin pumping out thyroxine. The hormone controls the disappearance of the tail and the development of limbs and other adult features. In mammals, thyroxine helps regulate metabolism.

Important insect hormones

Growth in arthropods such as insects involves the shedding of the hard outer skin, or exoskeleton. This process, called molting, is under close hormonal control. The brain releases a peptide hormone, PTTH, that stimulates the prothoracic glands. They lie at the front of the thorax. These glands make and secrete ecdysone, a steroid hormone that controls the shedding of the exoskeleton. Another hormone, juvenile hormone (JH), shuts down production of ecdysone when

IN FOCUS

The importance of metabolism

Growth depends on how quickly food can be converted into energy that can be used to make new cells. The conversion of food into energy is called metabolism. The metabolism of all vertebrates is regulated by hormones. The growth hormone somatotropin enables the metabolism of fats, which provide the body with the energy needed for growth. Somatotropin also stimulates the production of the other growth hormones. Many of these are produced in the liver and control the breakdown of proteins and carbohydrates. In humans, the pancreas and the adrenal cortex, among other glands, produce hormones that control the metabolism of glucose (a type of sugar).

Not all invertebrates have metabolic hormones, but insects do. Beside an insect's brain is a paired structure called the corpora cardiaca. This structure secretes a hormone that controls the metabolism of trehalose, a type of sugar that occurs in insect bodies.

molting is complete; the old exoskeleton has been shed, the animal has swelled a little, and a new, larger exoskeleton covers the body. Ecdysone functions only when the insect is growing. Adult insects do not molt; their prothoracic glands degenerate, and ecdysone production stops. JH is produced by glands called the corpora allata. They lie on either side of the esophagus near the brain.

Insects from several groups undergo the process of complete metamorphosis; a larval insect, such as a caterpillar, changes into an adult (in this case, a butterfly) through an immobile pupal stage. This process is regulated by JH. If lots of JH is present when the insect molts, it will pass to another larval stage. As JH levels drop, the larva molts into a pupa. A complete absence of JH allows the pupa to transform itself into the adult insect.

▼ *The process by which tadpoles develop into frogs is called metamorphosis. This change is controlled by the hormone thyroxine, which is produced by the tadpole's thyroid gland.*

Homeostasis

omeostasis is the regulation of the body's internal environment, features such as salt concentration, water levels, and temperature. Hormones are an important part of homeostasis. For example, they strictly regulate the concentrations of ions (charged particles) in the blood or in the extracellular matrix (the spaces between cells). Calcium, potassium, sodium, and phosphorus ions are the primary targets of this hormonal control system.

Birds, reptiles, amphibians, and fish have paired glands called pharyngeal ultimobranchial glands that produce and secrete the peptide hormone calcitonin. This controls the levels of calcium. Bony fish also have a unique structure called the stannius gland in the kidneys. The stannius gland produces a peptide hormone called hypocalcin, a potent calcium regulator.

Mammals produce calcitonin in C-cells, which lie in the thyroid gland. Calcitonin reduces levels of calcium in the blood in

▼ THYROID FEEDBACK
Human

This diagram shows the way in which thyroid hormone is regulated using a system called a negative feedback loop. When the level of thyroid hormone in the blood is higher than normal, the production of thyroid-stimulating hormone is inhibited. Thus, less thyroid hormone is produced.

Water regulators

The posterior pituitary glands of vertebrates produce a crucial peptide called antidiuretic hormone, or ADH. A diuretic speeds up the process of urine production. ADH regulates the amount of water in the body by acting on the kidneys. This allows the body to regulate urine output. Many insects produce an antidiuretic hormone similar to that of vertebrates.

two ways. First, the calcitonin suppresses the action of cells called osteoclasts that digest bone, releasing calcium and phosphorus into the blood. Second, the hormone increases the removal of calcium and phosphorus from the urine by acting on the kidneys.

The parathyroids and adrenals

Most vertebrates (though not fish) have a parathyroid gland, which consists of four pea-sized nodules attached to the back of the thyroid. Parathyroid hormone is the most important controller of concentrations of calcium and phosphorus in the body; levels of these ions must remain within very narrow ranges; otherwise, muscles and the nervous system cease to function correctly. Parathyroid hormone works in the opposite way to calcitonin, with the additional function of increasing the absorption of calcium from the small intestine. Parathyroid hormone is released in response to low levels of calcium and phosphate in the blood. It brings levels of these ions back to within normal range.

Humans have a pair of adrenal glands embedded in a wad of fat in front of the kidneys. One part of each, the adrenal cortex, is crucial for maintaining the correct concentration of potassium and sodium in the fluid that surrounds body cells. The hormones that achieve this are called mineralocorticoids.

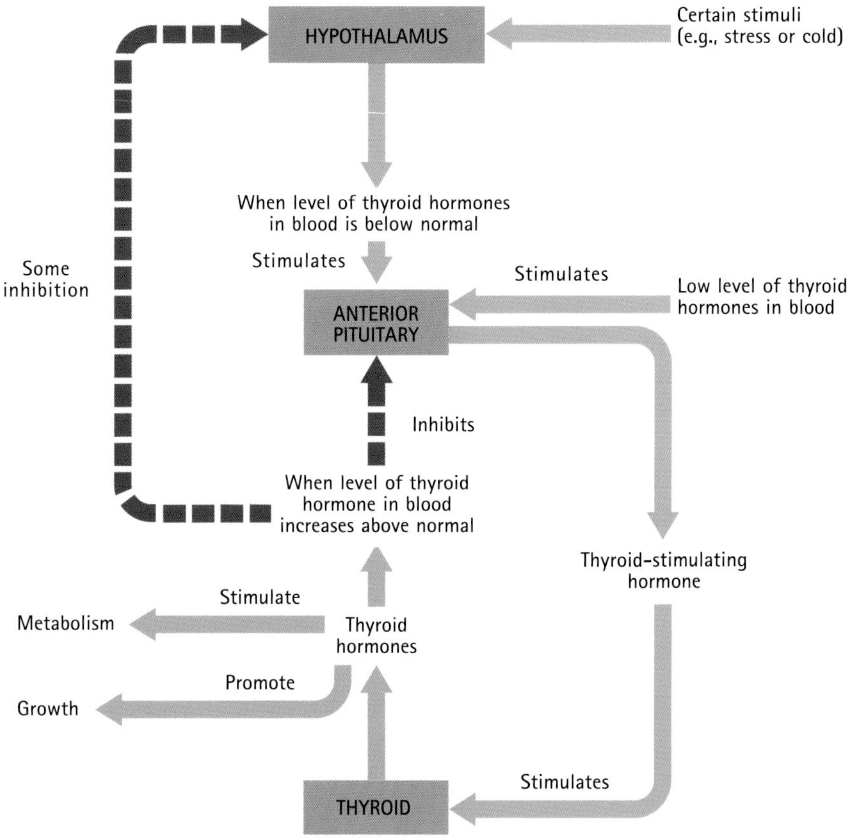

HYPOTHALAMUS

Certain stimuli (e.g., stress or cold)

When level of thyroid hormones in blood is below normal

Stimulates

Some inhibition

Stimulates

Low level of thyroid hormones in blood

ANTERIOR PITUITARY

Inhibits

When level of thyroid hormone in blood increases above normal

Stimulate

Metabolism

Thyroid hormones

Promote

Growth

Thyroid-stimulating hormone

THYROID

Stimulates

Reproduction

The timing of reproductive development in all vertebrates and most invertebrates is under hormonal control. Similar hormones that control sexual maturation occur in many different animal groups. For example, annelid worms have gonadotropin hormones very similar to those found in humans.

All animals have some endocrine control of reproduction. Worms of many groups have tissues in their brains that secrete hormones to control the development of gonads (sex organs). Nereidine is a hormone that controls both growth and the timing of sexual development in annelids. Other types of worms have hormones that control the timing of egg maturation.

Neurohemal gonadotropic hormones are responsible for the timing and development of sexual characteristics and sexual maturity in a variety of invertebrates, including mollusks. Glands in the eyestalks of crustaceans produce gonadotropins. Crustaceans' androgenic gland make peptide hormones that lead to the development of male sex organs; degeneration of this gland leads to the development of female sex organs. In insects, an endocrine gland in the brain produces a hormone that stimulates the manufacture of vitellogenin, a protein used by the insect's developing oocytes (eggs) to make egg proteins.

Reproduction in vertebrates

Lancelets are marine animals that are not vertebrates but look very like their ancient ancestors. Lancelets have glands similar to the vertebrate pituitary gland. The glands produce the steroid luteinizing hormone (LH), which stimulates the production of sperm in male lancelets. LH is important in vertebrates, too. It is made and released by the pituitary gland. LH is a gonadotropin; it stimulates the development of the gonads, or sex organs, and leads to the production of the sex hormones testosterone, in males, and estrogen, in females.

Other vertebrate gonadotropins include androgens and progesterone. In mammals, for example, progesterone is essential for helping embryonic young survive in the uterus.

▼ ESTROGEN CYCLE
Mammal

The release of eggs in mammals is called ovulation. The pituitary gland secretes follicle-stimulating hormone. This stimulates the ovaries to release a hormone called estrogen, which controls the development of an egg follicle. The pituitary gland also releases LH, which causes an egg follicle to break open, releasing the egg into the fallopian tube. There, it may be fertilized.

GENETICS

Fungal hormones

All multicellular organisms have reproductive hormones. Even fungi have them. Many fungi go through a complex "alternation of generations" during their life cycle, with an asexual (budding) stage followed by a sexual stage that involves fertilization by sperm. During the asexual stage, fungi are haploid: they contain only one set of genes. To move into the sexual stage, two haploid cells must unite to form a diploid cell—a cell with a full, double set of genes, as found in the body cells of most other animals, including humans. Fungi accomplish this change using hormones. They produce peptide hormones that ensure attraction and fusion of different haploid cells.

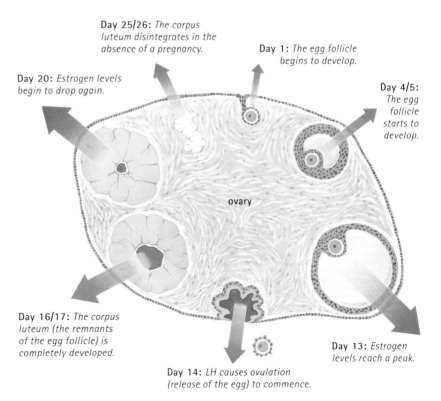

Day 25/26: *The corpus luteum disintegrates in the absence of a pregnancy.*

Day 1: *The egg follicle begins to develop.*

Day 20: *Estrogen levels begin to drop again.*

Day 4/5: *The egg follicle starts to develop.*

ovary

Day 16/17: *The corpus luteum (the remnants of the egg follicle) is completely developed.*

Day 13: *Estrogen levels reach a peak.*

Day 14: *LH causes ovulation (release of the egg) to commence.*

Vertebrate endocrine systems

The human endocrine system is very similar to that of other mammals. It also has much in common with the endocrine systems of non-mammalian vertebrates.

The basic structure of the endocrine system is similar in most vertebrates. In humans and most other vertebrates, the hypothalamus is located in the center of the base of the brain. The hypothalamus is a communication gateway, linking the nervous and endocrine systems. It contains cells that secrete chemicals called neurohormones. Some neurohormones trigger further hormone release, but others are inhibitory and prevent the secretion of other hormones. The hypothalamus receives signals from the body, then uses its neuro-hormones to signal to the next important gland—the pituitary gland.

The pituitary—the master gland

When it receives a neurosecretory signal from the hypothalamus, the pituitary gland releases one of its hormones. This may act on tissues directly or, more usually, may activate the production and release of other hormones from other glands. The pituitary is made up of two parts. The anterior (front) pituitary is like other glands, and secretes peptide hormones. The posterior (hind) pituitary is different. An extension of the hypothalamus, it consists mostly of axons (long extensions) of nerve cells originating in that part of the brain. All neurohormones from the hypothalamus go directly to the anterior pituitary and control all of its hormonal secretions.

Though it is controlled by the hypothalamus, the pituitary gland is often called the "master" gland. The pituitary synthesizes and releases a wide range of hormones, which influence all cells and most life processes in the body. The growth hormone, somatotropin, which targets liver and fat cells, originates in the pituitary. Somatotropin is crucial in stimulating the production of some other hormones needed for growth that are made by the liver and in other tissues. As growth requires energy, somatotropin is key in the metabolism of proteins, fats, and carbohydrates such as sugars. The timing and control of growth hormone release are regulated by a peptide hormone released from the hypothalamus, GHRH. GHRH also controls the release of a hormone that inhibits growth, called somatostatin, which is released from the hypothalamus, the pancreas, and several other endocrine organs.

Under orders from the hypothalamus, the anterior pituitary also produces thyroid-stimulating hormone, or TSH. TSH targets the thyroid gland. It causes proteins in the

pituitary gland

thyroid gland

adrenal glands

kidneys

The ovaries. Males have testes that lie outside the body cavity.

hypothalamus

parathyroids

pancreas

270

thyroid gland to be changed into the hormones T3 and T4. These hormones are involved in regulating the body's metabolic rate, growth and development, calcium levels, and the onset of puberty in humans.

Other pituitary hormones

Luteinizing hormone (LH) and follicle-stimulating hormone (FSH) are gonadotropins that stimulate the development of the sex organs and their hormones. LH and FSH are released by the anterior pituitary and are essential for reproduction in vertebrates. Prolactin is another pituitary hormone. In mammals, prolactin triggers mammary gland development and stimulates milk production during pregnancy and after birth.

Oxytocin is a peptide hormone formed in the hypothalamus. It passes to the posterior pituitary gland before moving into the bloodstream. Oxytocin stimulates the release of milk from the mammary glands of lactating female mammals. It also stimulates uterine contractions during childbirth, and may enhance bonding between a mother and her newborn young. In males, oxytocin enhances sperm mobility to increase the chances of fertilization. Although only mammals produce milk, many other vertebrates secrete oxytocin-like peptides. They stimulate the muscular contractions needed for laying eggs.

Another hormone produced in the pituitary is ACTH (adrenocorticotropic hormone), which acts on the cortex of the adrenal glands. The hypothalamus initiates the production of

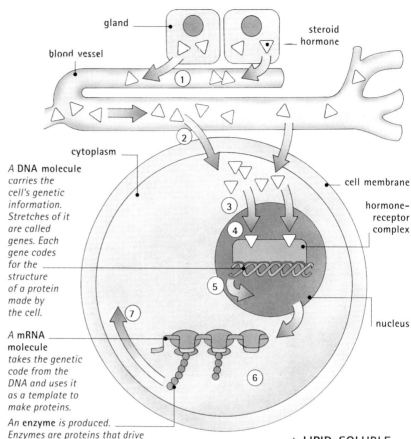

ACTH in response to stress. ACTH stimulates the adrenal glands to produce glucocorticoid hormones, such as cortisol.

The adrenal glands

In mammals, each adrenal gland has two parts. The inner medulla produces hormones such as catecholamines, epinephrine, and

▲ LIPID-SOLUBLE HORMONES

Steroid hormones, such as glucocorticoids and mineralocorticoids, are lipid-soluble. This characteristic allows them to diffuse through cell walls to bond with receptors within cells. A steroid hormone is secreted by a gland into the bloodstream (1) to a target cell. It then diffuses across the cell membrane (2), passing through the cytoplasm (3) to bind with a receptor at the nucleus (4). The hormone-receptor complex causes certain parts of the cell's DNA to be expressed (5). This leads to the production of enzymes (6) that alter cellular activity (7).

What controls endocrine activity?

The effect of a hormone depends largely on its concentration in the bloodstream at any one time. The amount of hormone encountered by a cell is influenced by three factors. The first is the rate of production of the hormone. This is controlled by feedback loops. As hormone levels begin to rise, mechanisms are triggered that cause them to fall. If there is too little of a hormone in the bloodstream, more will be produced.

The second factor affecting a cell's exposure to a hormone is its rate of delivery—a high blood flow delivers more hormone molecules than a low blood flow. The third factor is the hormone's "half-life." This is the rate of breakdown and elimination of the hormone. If a hormone is degraded rapidly and has a short half-life, its levels will drop rapidly. However, some hormones have long half-lives and can have effects long after secretion has stopped.

norepinephrine. It is surrounded by the cortex, which secretes two types of hormones: glucocorticoids and mineralocorticoids.

Catecholamine production is stimulated by stress: danger, fear, anger, hypoglycemia (low levels of blood sugar), exercise, and trauma. Epinephrine is another adrenal medullary hormone that "kicks in" under stressful conditions. The release of catecholamines has a dramatic effect on the body, increasing the heart, metabolic, and breathing rates, and stimulating the burning of fat to yield extra energy. Catecholamines also inhibit functions that are not essential at times of stress, such as digestion. Catecholamines break down rapidly after they are released.

Glucocorticoids and mineralocorticoids, the hormones produced by the adrenal cortex, are steroids; like all steroids, they are made from cholesterol. Glucocorticoids are crucial for regulating the body's metabolism of glucose, a kind of sugar. Every body cell contains receptors for glucocorticoids. Glucocorticoids also have strong anti-inflammation properties (they help reduce swellings). Cortisol, also known as hydrocortisone, is a well-known

glucocorticoid that is a commonly used anti-inflammatory drug. Mineralocorticoids are hormones that are vital for regulating the concentrations of potassium and sodium in the fluid that surrounds body cells.

IN FOCUS

Digestive hormones

Digestion is a complex process, closely regulated by the nervous and endocrine systems. Some of the gastrointestinal (GI) hormones that control digestion are produced in the brain, but many others are made and secreted by the digestive system itself. Important GI hormones include gastrin, which is secreted by the stomach and regulates gastric acid secretions; cholecystokinin, which is produced by the small intestine and stimulates the production of enzymes and bile in the pancreas; and secretin, another small intestine hormone that leads to the production of digestive fluids in the pancreas and liver.

▼ BLOOD SUGAR
Vertebrate

The level of glucose, a sugar, in the blood is regulated by two hormones produced by the pancreas: insulin and glucagon. When the amount of glucose in the blood gets too high, insulin acts to reduce the level. Glucagon does the reverse, causing the level of glucose to increase when it gets too low.

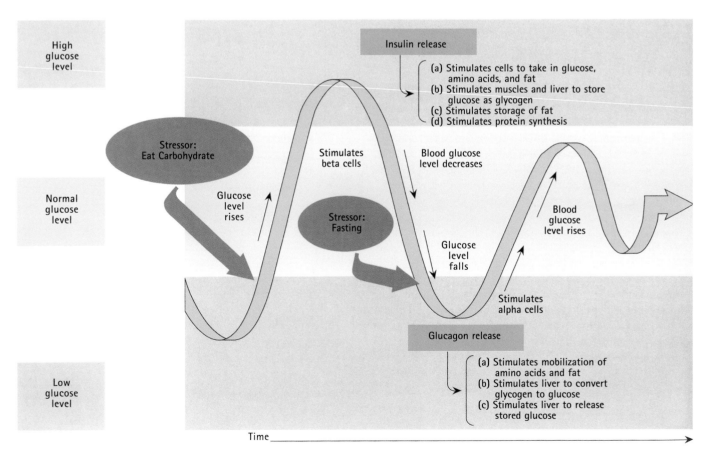

Hormones of the pancreas

The human pancreas is unusual in that it has both endocrine and exocrine functions. The pancreas is located next to the entrance to the small intestine. The endocrine part of the pancreas contains small clusters of cells called the islets of Langerhans, which produce several hormones. One of the hormones—somatostatin—inhibits growth; the other two are glucagon and insulin.

Insulin is a peptide hormone vital for the metabolism of carbohydrates and fats, and involved in the breakdown of proteins and minerals. Glucose is produced by the digestion of carbohydrates in the small intestine. When glucose moves into the bloodstream, insulin is released. Insulin enables cells in muscle and other tissues to take in glucose and use it as a source of energy. Insulin is also required to make the liver store glucose in the form of glycogen, so insulin lowers glucose concentrations in blood by allowing its uptake by cells around the body. As blood glucose levels fall, insulin production stops. Cells then cannot take up glucose, so they switch to alternative sources of energy, such as fatty acids. When blood sugar levels increase, insulin is again produced.

Glucagon works in the opposite way to insulin. This peptide hormone increases blood glucose levels by making the liver break down its glycogen reserves. Glucagon secretion is stopped by high levels of glucose in the blood.

Insulin and glucagon have similar functions in other vertebrates. All tetrapods (four-legged vertebrates) have discrete pancreatic glands. Fish, however, have pancreatic cells distributed widely through their digestive tract.

The pineal gland

The pineal gland, or epiphysis, is a small gland located in the brain. Shaped like a pinecone, the pineal gland secretes the hormone melatonin, which communicates information about light conditions in the environment to the body. Light that enters the eye is transmitted as a signal to the hypothalamus. This communicates the information via the spinal cord to the pineal gland.

The nerve signal triggers the production of melatonin from another chemical, serotonin. Melatonin production is lowest during the

The third eye

In amphibians and in many species of reptiles, the pineal gland is photosensitive; it is directly sensitive to light, although it cannot form images. It is often called the parietal eye, or third eye. The parietal eye is particularly well developed in lizardlike reptiles called tuataras, which live only on islands off New Zealand. They have a hole in the skull called the parietal foramen that allows light to reach the gland.

Photosensitivity allows the pineal gland to regulate biological rhythms by nervous signals to other regions of the brain. Direct photosensitivity has been lost in mammals, but the pineal gland remains important in regulating these rhythms. It is supplied with information on light levels by nervous signals originating in the visual cortex of the brain.

day when light levels are at their maximum. Melatonin output reaches a peak at night. The hormone helps regulate the body's circadian (daily) rhythms, including sleep–wake cycles and levels of daily activity. It also affects reproduction by inhibiting the production of the gonadotropins that lead to the development of sex organs and hormones. Irregularities in melatonin output cause insomnia and are also linked to seasonal affective disorder (wintertime depression).

▼ *Milk production is triggered by prolactin, which is released by the pituitary gland. Suckling causes the release of further hormones that maintain milk output.*

Exocrine system

▶ **EXOCRINE GLANDS**
The range of exocrine glands that occur in mammals.

simple tubular

simple coiled tubular

simple branched tubular

simple branched tubular (with ducts)

simple acinar

simple branched acinar

simple branched acinar

CONNECTIONS

COMPARE the pheromone glands of an *ANT* with those of a *LION*.

COMPARE the poison production mechanisms of a cane toad with those of a *TARANTULA* and of a *SCORPION*.

Unlike endocrine glands, most of which release their hormones directly into the blood, the glands of the exocrine system release the chemicals they produce onto the surface of the body or into cavities inside it, such as the gut. Most exocrine glands secrete their products through ducts. They are generally simple tubular structures through which secretions flow, though some exocrine glands contain branching ducts of various sizes. Exocrine glands are classified by the type of secretion they release. Serous glands have cells that produce and release serous, or clear, watery secretions, such as tears. Mucus glands produce heavier mucus secretions. Mixed glands, such as the salivary glands, contain both kinds of cells, and their secretions are a combination of both.

Salivary glands

The salivary glands secrete chemicals that begin the digestion of food. In humans, the numerous salivary glands are scattered about the mouth and on the lips, on the insides of the cheeks, and on the tongue. However, there are three major pairs of these glands. The parotids are the largest; they are located in front of and below the ears. The sublingual glands are on the floor of the mouth under the tongue. The submaxillary glands are located under the jaw. The salivary glands release saliva, a mixture of water, ions, mucus, and digestive enzymes.

Saliva has many important functions. It aids in the digestion of food, since it contains amylase, an enzyme that breaks down starches

into sugars; and it helps bind food into a slippery bundle that is easily swallowed. Saliva also helps keep the body cool. Dogs and some other animals pant when they are hot; the evaporation of saliva cools blood passing through vessels close to the tongue's surface.

The sweat glands

Maintaining a constant temperature is essential for warm-blooded animals, so most also lose heat through the evaporation of sweat. Perspiration is released from sweat glands, traveling through ducts and out of pores to the surface of the skin. In humans, overstimulated

IN FOCUS

Territory markers

Vertebrates use pheromones as sexual attractants and for territorial marking. The females of many mammals produce pheromones during estrus to attract mates. Some male mammals use pheromone-laced urine to mark their territory. For example, the dik-dik (a type of African antelope) has scent glands at the inner corners of its eyes. By touching a gland to a twig or branch, the dik-dik marks the boundaries of its territory. Male sungazer lizards also use pheromones; theirs are released through pores on the underside of the thighs.

nerves also trigger the release of sweat. Sweat glands are connected to nerves of the sympathetic nervous system, so they respond to emotional states, such as fear. Fear-induced sweating is stimulated by the release of epinephrine from the adrenal glands.

The lachrymal glands

Humans and most other vertebrates have lachrymal glands—better known as tear ducts. They produce serous secretions that maintain moisture on the eyes. The lachrymal glands are almond-shaped and are located at the outer corner of each eye. A conjunctival mucus gland adds a little mucus to tears.

Human eyes are constantly washed by a film of tears, which keeps them lubricated and clean. The film has three layers—an oily layer, a watery layer, and a mucus layer. All are essential for the health of the eye. Blinking helps maintain an even coating of tears over the whole of the eyes exposed to air.

Attack and defense

Animals use exocrine secretions for a range of purposes: for attracting mates, for establishing territories, in defense, or for hunting. Many insects, including female moths, produce and release chemicals called pheromones to attract mates. Neurohormones released by an insect's brain cause specialized body tissues to release pheromones, which are also important for regulating group behavior among social insects, such as ants.

Exocrine secretions may be used for hunting prey. Some snakes, for example, use venom to immobilize prey. Cobras have venom glands all along their jaw and fangs. The venom is delivered through the hollow needlelike fangs. Many other animals use exocrine secretions for defense. The red lionfish has glands at the base of each of its sharp spines. When an attacker comes into contact with a spine, the gland releases a potentially lethal chemical. Bombardier beetles have a pair of glands near their anus that produce a spray containing a cocktail of hot, corrosive, and toxic chemicals. Skunks, too, are famous for their chemical defenses. The distasteful odor of a threatened skunk is produced by secretions from the animal's anal glands.

NATALIE GOLDSTEIN

Marvelous milk

One of the most important exocrine secretions is milk. Found only in mammals, milk is secreted by the female's mammary glands. It is full of nutrients that help a young mammal grow quickly. Milk varies greatly in composition between species. For example, species with young that need to lay down a thick layer of blubber as soon as possible after birth, such as seals, have milk that is extremely rich in fat. Seal milk contains up to 60 percent fat.

▼ Mammary glands

In humans, milk emerges from the nipple through several ducts. In ungulates (hoofed mammals), milk is discharged through a single duct.

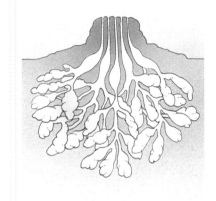

Human

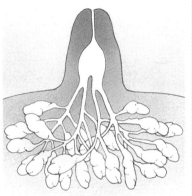

Ungulate

FURTHER READING AND RESEARCH

Silverthorn, Dee. 1998. *Human Physiology: An Integrated Approach*. Prentice Hall: Upper Saddle River, NJ.

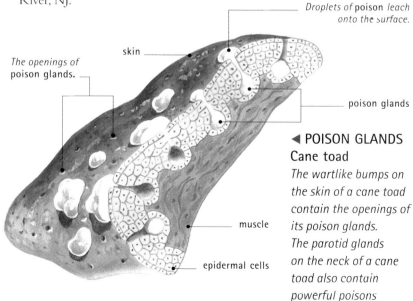

Droplets of **poison** *leach onto the surface.*

skin

The openings of poison glands.

poison glands

muscle

epidermal cells

◄ POISON GLANDS
Cane toad

The wartlike bumps on the skin of a cane toad contain the openings of its poison glands. The parotid glands on the neck of a cane toad also contain powerful poisons that deter predators.

Fern

DIVISION: **Pterophyta** ORDERS: **up to 20**
FAMILIES: **about 40**

The ferns are a group of non-flowering plants that need to live in wet places to reproduce efficiently. This requirement restricts ferns to shady environments that do not easily dry out. Ferns have evolved the ability to flourish in shady locations, such as the lower levels of tropical rain forests. There, the amount of light reaching them is too low for most other plants.

Anatomy and taxonomy

All life-forms are classified in groups of relatively closely related species. The classification is based mainly on shared anatomical features, which usually (but not always) indicate that the members of a group have the same ancestry. Thus the classification shows how life-forms are related to each other. Scientists can also compare the anatomy of living organisms and fossil forms preserved in rocks of known age. This can indicate how long the various groups have existed. These studies show that ferns are an ancient group that evolved around 400 million years ago—long before the first flowering plants.

- **Plants** All true plants are multicellular organisms. Unlike fungi or animals, most plants are able to make their own food from simple chemicals in the air and soil. They absorb carbon dioxide gas and water, and use the energy of sunlight to convert these chemicals into carbohydrates. These are further converted into starches, cellulose, and other materials. Plants absorb the sun's energy using a green pigment called chlorophyll, which is responsible for the green color of most plants. Typical plants also use soil nitrates, phosphates, and other substances to build proteins and other compounds essential for growth. A few plants live as parasites by stealing the food made by other plants.

- **Vascular plants** All plants need water. Some plants, such as mosses, absorb water directly into the tissues that need it, so all their tissues must be in close contact with water. Other plants have a system of tubes that carry water up through the roots, stems, and leaves. This system is called a vascular system, and it enables plants to grow much taller, so their upper parts are not in direct contact with water. All seed-bearing and flowering plants have vascular systems and are known as higher vascular plants. Some groups of plants that have spores instead of seeds also have vascular systems; they are known as lower vascular plants.

- **Lower vascular plants** The plant life of Earth is dominated by plants that produce seeds. Each seed is created by the fusion of a male pollen cell with a female

This family tree shows all the major groups to which ferns belong.

Plants
KINGDOM Plantae

Nonvascular plants

Vascular plants

Lower vascular plants

Seed-bearing plants

Club mosses
DIVISION Lycophyta

Whisk ferns
DIVISION Psilophyta

Ferns
DIVISION Pteridophyta

Horsetails
DIVISION Sphenophyta

Eusporangiate ferns
ORDER Ophioglossales

Leptosporangiate ferns
ORDER Filicales

FAMILY Ophioglossaceae
2 genera, including *Ophioglossum*
adder's tongue ferns

FAMILY Polypodiaceae
47 genera, including
Platycerium stag's horn ferns

FAMILY Hymenophyllaceae
11 genera, including
Hymenophyllum filmy ferns

FAMILY Dennstaedtiaceae
14 genera, including
brackens *Pteridium*

FAMILY Woodsiaceae
21 genera, including
Athyrium lady ferns

ovum. The pollen is carried on the wind, or transported by animals such as insects or birds. Lower vascular plants such as ferns do not produce seeds but rely on male sex cells swimming through water to reach the female cells. The fusion of the two creates a fertile cell. This grows into a structure called a sporophyte, which produces spores. Spores are tough but light and can be blown to new locations. Each spore can grow into a form called a gametophyte, which produces more male and female cells.

<table>
<tr><td rowspan="6" style="writing-mode:vertical">ANATOMY HIGHLIGHTS</td></tr>
</table>

EXTERNAL ANATOMY Ferns are mostly small plants with an underground rhizome from which roots and stalks arise. Fronds of leaves (pinnae and pinnules) grow from the stalks. *See pages 278–280.*

INTERNAL ANATOMY Water, sugar solution, and other chemicals move through vascular tubes. An outer layer of cells (the epidermis) protects the leaves and stalks. *See pages 281–282.*

REPRODUCTIVE SYSTEM Ferns reproduce by sexual and asexual reproduction. Male and female sex cells are borne on spores released from the leaves. The male sex cells swim to female cells to fertilize them. Hence, ferns thrive only in damp habitats. *See pages 283–285.*

▲ *A lady fern. This deciduous fern forms dense clumps in meadows, moist woodland, and open thickets, and its delicate lacy fronds may grow up to 5 feet (1.5 m) in height.*

● **Ferns** The ferns are the most numerous of the lower vascular plants, with 10,000 to 15,000 species. They are very variable in form, but most grow from underground stems called rhizomes and have upright green leaves that are usually divided into smaller leaflets. Other types of lower vascular plants have different structures. The stems of club mosses are covered with small spiny leaves; horsetails have whorls of very small scaly leaves that sprout from joints in their stems; and whisk ferns have no leaves at all.

● **Fern families** There are up to 20 orders of ferns (botanists disagree on the number), divided into roughly 40 families. They include plants ranging from the very small free-floating aquatic ferns of the family Azollaceae to the very large tree ferns of the family Dicksoniaceae. Others include the delicate filmy ferns of the family Hymenophyllaceae, which have leaves only one cell thick; and the much sturdier, more typical ferns such as those of the family Polypodiaceae, many of which grow in temperate regions such as North America and Europe.

External anatomy

COMPARE the big, compound leaves of a fern with the small, simple leaves of an **APPLE TREE**. The form of the leaves is very different, but both have the same function.

A typical fern has a clump of long, graceful green leaves, or fronds, sprouting from near the ground in a shady, damp place. Some ferns, such as the hart's-tongue ferns, have simple straplike leaves, but most species have feathery compound leaves. The main leaf stem, or petiole, extends into a central rib called a rachis. This carries a diminishing series of leaflets, or pinnae. The shape of each pinna varies according to the species, but it is usually made up of a series of even smaller pinnules. Together they form a leaf with a large surface area. This arrangement helps the plant gather

▶ ROOTS
AND LEAVES
Lady fern
The lady fern is a typical fern with creeping rhizomes and upright, compound leaves. Each leaf has many small leaflets, each of which has even smaller pinnules.

The **pinnae**, *or leaflets, are made up of even smaller* **pinnules**. *The shape of a pinna replicates that of the whole leaf.*

pinnules

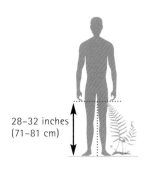

28–32 inches
(71–81 cm)

pinnae

Water is drawn up the **rachis** *and into the leaves for use in photosynthesis.*

petiole

The young **leaves** *are tightly curled into "fiddleheads," which gradually unfurl as they grow.*

The **rhizome** *stores food resources. It grows horizontally below ground level, sprouting new leaves at intervals.*

The **roots** *gather water from the soil, together with dissolved nitrates, phosphates, and other essential plant nutrients.*

the greatest possible amount of light in shady conditions, and it also allows water to drain through the leaf easily, helping to prevent damage to the leaf and its stalk.

In most ferns all the leaves on a plant look alike, apart from their size. Typical ferns bear spores beneath all their mature leaves, but some ferns produce specialized spore-carrying fronds. The royal fern has been called the "flowering fern," because in summer its leaf tips bear flowerlike sprays of highly modified, spore-producing pinnae.

Creeping rhizomes

The green leaves of a fern generally grow from a thick, fleshy underground stem called a rhizome, which acts as an energy store during the winter when the fern cannot make food. The rhizomes have slender roots that gather water and vital plant nutrients from the soil. Fern rhizomes can also grow through the ground, sprouting new clumps of leaves that look like separate plants. Some ferns, such as bracken (genus *Pteridium*), can cover large areas like this, even overwhelming other plants.

Although most ferns are fairly small plants, with leaves sprouting directly from the underground rhizome, there are plenty of exceptions. The most spectacular are the tree

COMPARATIVE ANATOMY

Hairy horsetails

Other lower vascular plants are quite unlike typical ferns. The most common and conspicuous are horsetails, which have jointed stems that sprout whorls of smaller, hairlike branching stems like bottle brushes. They have tiny leaves that are just toothed rings surrounding each stem joint. Like ferns, hairy horsetails sprout from creeping rhizomes and can grow vigorously. Horsetails were very common during the Carboniferous period, 350 million to 280 million years ago, along with ferns and club mosses. Some horsetails grew to the size of modern trees. Paleontologists know this because the compressed, fossilized remains of horsetails have survived as coal in rocks of Carboniferous age.

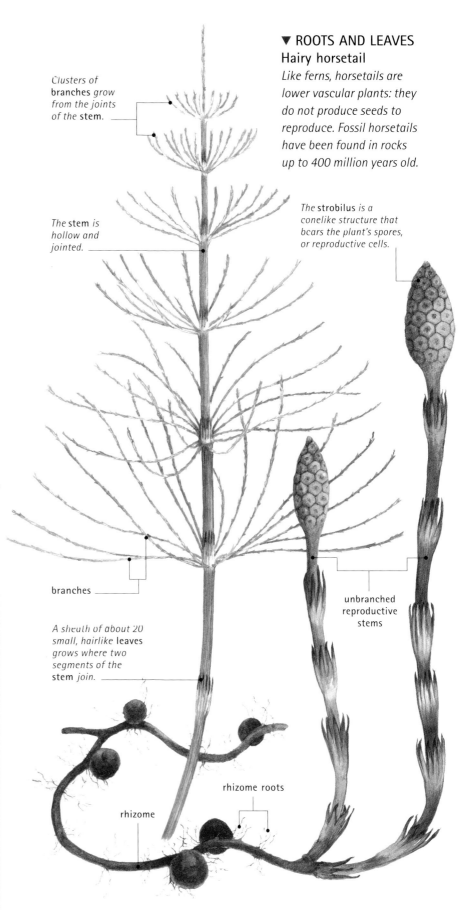

▼ ROOTS AND LEAVES
Hairy horsetail
Like ferns, horsetails are lower vascular plants: they do not produce seeds to reproduce. Fossil horsetails have been found in rocks up to 400 million years old.

Clusters of branches grow from the joints of the stem.

The stem is hollow and jointed.

The strobilus is a conelike structure that bears the plant's spores, or reproductive cells.

branches

A sheath of about 20 small, hairlike leaves grows where two segments of the stem join.

unbranched reproductive stems

rhizome roots

rhizome

279

ferns, which can grow to 60 feet (20 m) tall. These ferns look like palm trees, with a thick trunk and a whorl of leaves at the top. The trunks of tree ferns are so big and strong that they are even used to build houses.

Most tree ferns grow in tropical forests, but a few species flourish in the cool temperate rain forests of southeastern Australia, New Zealand, and the southern Pacific islands. Some of these temperate species, including *Cyathea* and *Dicksonia,* are grown as ornamental plants.

Tree ferns can grow only from the top of the plant. They do not have side branches like many other plants. New fronds sprout from the center whorl above the bases of the old leaves, as in a palm tree. Gradually the woody leaf bases build up into a tall, strong trunk.

Unlike flowering plants, tree ferns do not form new woody tissue in the trunk as they grow (a process called secondary growth). Instead, as the tree fern grows taller, roots sprout from the trunk. The thick, fibrous mass of roots expands as the tree fern grows, and supports the increasing weight.

▼ *This tree fern has a strong, woody trunk. Many leaves grow from the top of the trunk, which grows a little taller each year.*

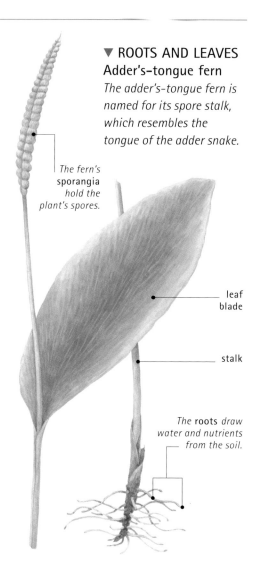

▼ ROOTS AND LEAVES
Adder's-tongue fern
The adder's-tongue fern is named for its spore stalk, which resembles the tongue of the adder snake.

The fern's sporangia hold the plant's spores.

leaf blade

stalk

*The **roots** draw water and nutrients from the soil.*

Some ferns grow on other trees. The stag's-horn fern roots in the bark of a tropical forest tree, high above the ground. It develops two roughly circular leaves that form a cup for gathering moist, nutritious leaf litter. It gets its name from its specialized spore-carrying fronds, which resemble deer antlers.

Water ferns

A few ferns live exclusively in the water of streams and pools. *Marsilea* has small, green, cloverlike floating leaves that sprout from a slender rhizome rooted in the bottom mud. By contrast *Azolla* is a tiny, free-floating plant that turns bright red in summer. It lives in partnership with microscopic cyanobacteria (formerly blue-green algae) that are able to gather vital nitrogen directly from the air. *Azolla* is so good at gathering nitrogen that it is grown in rice paddies. When *Azolla* decomposes it fertilizes the rice like a green manure.

Internal anatomy

CONNECTIONS

COMPARE the vascular bundles of a fern with the similar structures of a *MARSH GRASS*. The grass is a higher vascular plant, but its vascular system works in much the same way.

COMPARE the rhizome of a fern with the tuber of a *POTATO*. Both act as energy stores, but potato tubers grow at the ends of the plant's roots.

Ferns are relatively primitive plants, but they have a very complex structure that is the product of many millions of years of evolution. Ferns can survive in places too dark for most flowering plants. Some ferns, such as bracken, can even outcompete flowering plants on open, sunny hillsides.

Solar panels

One reason for this success is the efficiency of a fern's leaves. Each leaf of a typical fern expands into a broad sheet that presents as many cells as possible to the light and air. The delicate leaves of one group of species, filmy ferns, are just one cell thick, enabling every part of the leaf to gather light and vital gases. Most fern leaves have three cell layers: a strong skin, or epidermis, above and below, separated by a layer of cells called the mesophyll.

The light energy that falls on the leaf is absorbed by green chlorophyll, which is contained in microscopic structures called chloroplasts within the cells. The leaves also absorb carbon dioxide from the air. Meanwhile, water is drawn up from the roots through the plant's vascular system. The light energy promotes a complex chemical reaction between the carbon dioxide and water. The reaction turns carbon, hydrogen, and oxygen into a carbohydrate called glucose, which is a simple sugar. This is then carried off through a leaf vein network to other parts of the plant in a fluid called sap. This process of using light to make food is called photosynthesis.

Vascular bundles

Since ferns are able to photosynthesize in sunlight that is too faint for many plants, ferns thrive in many shady environments. However, ferns need a plentiful supply of water, which most gather from the soil. As with other groups of vascular plants, the water-gathering mechanism of ferns is powered by the leaves, which have small holes in their lower surfaces called stomata. The stomata release water into the air, and the leaf replaces it with water from

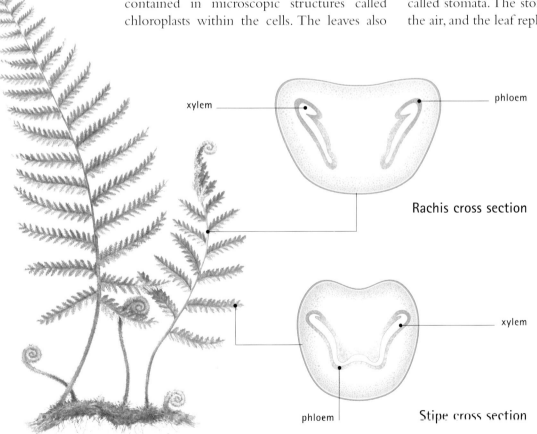

xylem

phloem

Rachis cross section

xylem

phloem

Stipe cross section

◀ VASCULAR BUNDLES
Generalized plan
The shape of the vascular bundle is not the same throughout a fern. In the rachis of the fern Onoclepsis *the vascular bundle is divided. However, the vascular bundle is W-shaped in the stipes.*

281

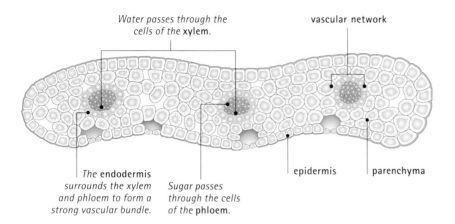

Water passes through the cells of the **xylem**.

vascular network

The **endodermis** surrounds the xylem and phloem to form a strong vascular bundle.

Sugar passes through the cells of the **phloem**.

epidermis

parenchyma

▲ LEAF CROSS SECTION
Bracken
Water is drawn along each xylem to be used in photosynthesis. Sugary sap passes along the phloem. Some of the sugar solution is converted to cellulose, the tough fiber that forms the walls of new cells. Sugar also fuels the fern's metabolic processes.

its leaf stem and rhizome. This in turn causes water to be drawn from the roots to the rhizome and stem, which replace it by absorbing water from the soil.

The water-gathering mechanism depends on an efficient plumbing network, or vascular system. In ferns this takes the form of one or more bundles of tubes that pass through the roots, rhizomes, stems, and leaves. At the center of each bundle is a group of tubular cells called tracheids. The tracheids carry water up from the roots and deliver it to the leaves for use in photosynthesis. The bundle of tracheid cells is called the xylem.

The bundles of vascular tissue within the rhizome are called the stele. Where the stele forms a continuous ring it is a called a siphono-stele. Where the siphonostele is not continuous it is called a dictyostele. In a dictyostele arrangement, the phloem surrounds bundles of xylem tissue. In a siphonostele arrangement, the ring of xylem tissue has phloem on both the outside and the inside.

IN FOCUS

Leaf growth

When fern leaves grow they unfurl from tightly coiled spirals that resemble the curly bit at the top of a violin, so they are sometimes called fiddleheads. The coiled growing tips consist of cells that keep multiplying by dividing in two. Once a section of leaf has straightened out, its cells stop dividing. The leaf tips of some tropical ferns may keep expanding in this way for years, until they are 10 feet (3 m) long. In cooler regions the fern foliage dies down each winter, and new fiddleheads grow from among the remains of the old leaves in spring.

EVOLUTION

Ancient ancestors

The strangest lower vascular plants are a small group of species called whisk ferns. They are branched green stems that sprout from rhizomes, and they have no true roots or leaves. Yet whisk ferns do have threadlike rhizoids on their rhizomes. The rhizoids contain microscopic fungi that live in partnership with the plants, helping them absorb vital nutrients from the soil. Astonishingly, exactly the same structures have been discovered in plant fossils that are 400 million years old. The fossilized plants look very like modern whisk ferns, indicating that whisk ferns may have lived on Earth for a very long time indeed.

Energy store
Some of the sugary sap flows down into the thick rhizome of the fern, where some of the sugar is converted into another carbohydrate called starch. The starch is stored in the cells of the rhizome. When necessary, the fern can use the starch as a source of energy by converting it back into sugar. This ability to convert starch to sugar is vital for ferns that live in climates that experience a cold winter such as North America and Europe; the energy store helps the plant survive winter and fuels the production of new leaves in the spring.

Reproductive system

CONNECTIONS

COMPARE the alternating generations of ferns with the alternating generations of *JELLYFISH*. The polyp stage of the jellyfish life cycle is quite unlike the swimming medusa stage, just as the fern gametophyte is quite unlike the leafy sporophyte.

Atypical flowering plant grows from a seed, which is formed when a male sex cell fertilizes a female sex cell. Ferns, however, have a different reproductive system. They do not produce seeds. Instead, each large fern plant produces thousands of microscopic, dustlike spores, which can grow into new plants (gametophytes) without being fertilized.

Yet the gametophytes are not like the parent ferns. They are small, flattened, dark green structures, anchored to wet ground by short rootlike rhizoids. The plants grow quickly and do not live long. They produce male and female sex cells, which fuse to create fertile embryos and grow into new leafy ferns.

CLOSE-UP

The gametophyte

The fern gametophyte is a fleshy, heart-shaped structure, usually about the size of a dime. The gametophyte grows close to the ground in wet places, anchored by hairlike rhizoids on its underside. Sperm-producing antheridia develop as small knobs among the rhizoids. Egg-producing archegonia also grow on the underside, usually near the notch of the heart shape.

► LIFE CYCLE
Capsules, or sporangia, containing sporocytes, grow under the leaves of mature ferns. When the sporangia are ripe, they break open, releasing the sporocytes and their spores. Each spore can grow into a gametophyte, which carries both male and female sex cells. When the sperm are released in water, they swim to fertilize a female sex cell that has been released from the archegonium. The fertilized cell, or zygote, develops into a young fern, and the cycle begins again. The large, heavy form of the fern that produces the gametophyte is called a sporophyte.

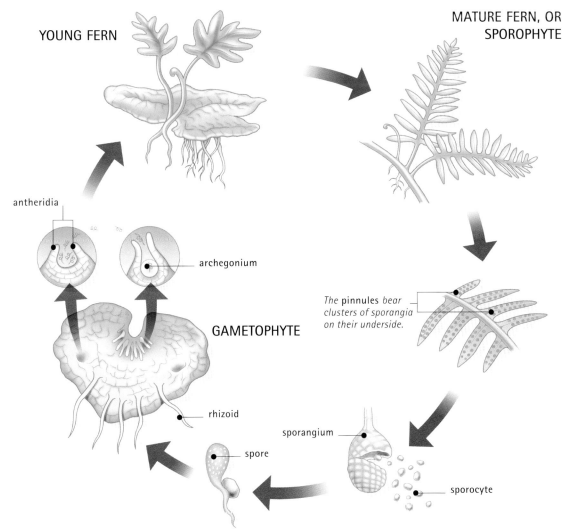

YOUNG FERN

MATURE FERN, OR SPOROPHYTE

antheridia

archegonium

GAMETOPHYTE

*The **pinnules** bear clusters of sporangia on their underside.*

rhizoid

spore

sporangium

sporocyte

283

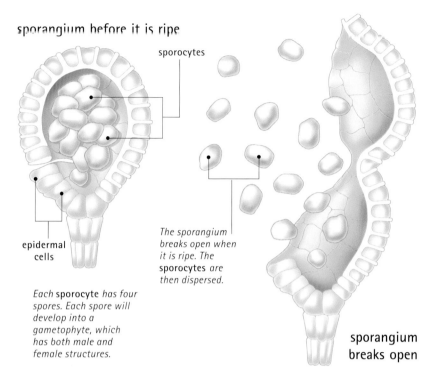

sporangium before it is ripe

sporocytes

epidermal
cells

Each **sporocyte** *has four spores. Each spore will develop into a gametophyte, which has both male and female structures.*

The sporangium breaks open when it is ripe. The **sporocytes** *are then dispersed.*

sporangium
breaks open

▶ SPORE DISPERSAL
Generalized fern
In most ferns, groups of sporangia develop on the underside of leaves. The sporangia contain the plant's sex cells. When a sporangium is ripe, it splits open and releases spores, which either fall to the ground beneath or are carried farther by a breeze. If a female cell is fertilized by a male cell, a zygote is created. If this happens in a damp place, the zygote may develop into a new fern.

As a result of this arrangement ferns occur in two alternating forms: as big, leafy, spore-producing sporophytes, and as small, flattened, sexual gametophytes. People recognize only the leafy sporophytes as ferns because the gametophytes are so small.

Other lower vascular plants reproduce by the same system. Club mosses and horsetails carry their spores in conelike structures on the end of shoots, and their gametophytes are tiny. The gametophytes of horsetails are only about the size of a pinhead.

Shedding spores
Some fern sporophytes, such as the royal fern, stag's-horn fern, and adder's-tongue fern, have spore-producing fronds. However, a typical fern, such as the lady fern, produces its spores from structures called sporangia beneath its leaves. The sporangia are often grouped together in clusters called sori, which may be protected by covers like tiny umbrellas.

Each sporangium is filled with sporocytes, or spore "mother cells," and each sporocyte produces four spores. A sporocyte is a diploid cell: that is, it has two full sets of chromosomes, which carry the genes controlling how the fern develops. But the spores that the sporocyte produces are haploid: each has just one full set of chromosomes. Male and female sex cells

are also haploid, and when a sperm fertilizes an egg their chromosomes pair up to make a fertile diploid cell.

When the spores are ripe, the sporangia pop open to toss them out. Some will fall to the ground under the fern, but if there is a breeze others may be carried farther away. Of the thousands of spores released, only a few may land in a damp place and grow into small, flattened, heart-shaped gametophytes.

Fertilization
Some lower vascular plants, such as quillwort, produce male and female spores that grow into male and female gametophytes. Some water ferns do the same, but the spores of all other ferns grow into gametophytes that carry both male and female structures on their underside.

The male structures are called antheridia, and they develop when the gametophyte is very young. Each antheridium produces 32 haploid sperm cells, which can swim through water by beating mobile flagella. The female structures, or archegonia, develop later; each contains a single haploid egg cell.

When the sperm cells are fully developed they leave the antheridia and swim away through the film of water on the ground or beneath the plant. They are microscopically small, so they need only a thin water film.

GENETICS

Back-up genes

Fern spores do not grow into complete ferns because each spore is haploid: it has only one set of the genes needed for the development of a complete fern. If some of the genes are damaged, the spore will not develop properly. A diploid cell, created when a sperm fertilizes an egg, has two sets of genes, so if one gene is faulty, the cell has a back-up copy. Since eggs and sperm from the same plant are likely to have the same faulty genes, sexual reproduction works best if sex cells come from different plants. The egg cells on each gametophyte often develop later than the sperm cells, so there is less chance of a male sex cell fertilizing a female cell from the same plant.

COMPARATIVE ANATOMY

Mosses

The leafy fern plants that people see growing in shady places are the sporophyte generation. But among mosses, which have a similar two-phase life cycle, the green plants that we see are the gametophyte generation. A moss produces egg and sperm cells, and these fuse to produce small sporophytes that sprout up from the plant rather like seed heads. The sporophytes then scatter spores that develop into new mosses.

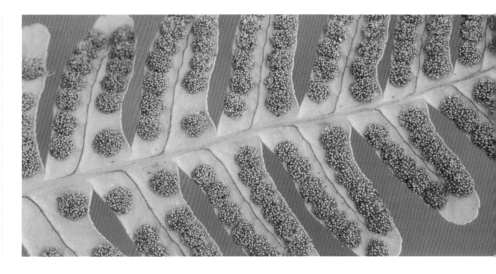

If there is no water they cannot escape; this is why ferns nearly always grow in damp places. If a mobile sperm cell manages to get into an archegonium, either on the same gametophyte or a neighboring one, it fuses with the egg cell inside to create a diploid cell, or zygote. This process is the equivalent of a seed in a flowering plant. The zygote sprouts roots, stems, and leaves, and develops into a young sporophyte, or fern. Meanwhile, the gametophyte shrivels and disappears.

Hanging on

Club mosses have a fernlike relative called spike moss, which produces two types of spores: male and female. Instead of growing into separate male and female gametophytes, like those of another relative, quillwort, the larger female spores stay on the plant that produced them. Most spike mosses live in wet places where the moisture enables sperm cells to swim over their leaves and fertilize the egg cells produced by the female spores. The zygotes then drop off, like seeds, and sprout into new spike mosses.

Vegetative reproduction

Ferns can also spread by extending their rhizomes underground. New leaves and roots grow from the extensions, and if the older sections of the rhizome die off, or the rhizome is cut, the new sections keep growing. This vegetative growth allows species such as bracken to spread well beyond the shady woodland habitats typical of most ferns.

New ferns that grow in this way are clones—they are genetically identical to the ferns from which they sprouted. If they are still physically attached to the parent plant they could even be considered the same plant, so a single fern plant may cover a huge area and live for hundreds of years.

Some types of ferns, such as *Asplenium bulbiferum,* are also able to reproduce by sprouting buds on their leaves. The buds turn into new plants that drop off the parent and take root. These plants are also clones of the parent plant because the process does not require the fusion of two sex cells.

JOHN WOODWARD

▲ *Each brown area on the underside of the pinnules of this fern is called a sorus. A sorus is a group of sporangia. Each sporangia is made up of many sporocytes.*

FURTHER READING AND RESEARCH

Weier, T. Elliot, C. Ralph Stocking, M. G. Barbour, and T. L. Rost. 1982. *Botany.* Wiley: New York.

EVOLUTION

Breaking away from water

The egg cells of a flowering plant are contained in structures called ovules at the base of each flower. The ovules may have evolved from gametophytes that stayed on the parent sporophyte, like those of some spike mosses. However, their sperm cells are carried in pollen grains, either on the wind or by animals, so they do not need to swim through water to reach the egg cells. This method of dispersal enables flowering plants to grow in places that are not always wet, unlike spike mosses and most ferns. It is one reason why flowering plants are so much more widespread than mosses.

Index